CAMBRIDGE LIBRARY COLLECTION

Books of enduring scholarly value

Life Sciences

Until the nineteenth century, the various subjects now known as the life sciences were regarded either as arcane studies which had little impact on ordinary daily life, or as a genteel hobby for the leisured classes. The increasing academic rigour and systematisation brought to the study of botany, zoology and other disciplines, and their adoption in university curricula, are reflected in the books reissued in this series.

Historical and Biographical Sketches of the Progress of Botany in England

Richard Pulteney (1730–1801) was a Leicestershire physician whose medical career suffered both from a lack of aristocratic patronage and from his dissenting religious background. However, his lifelong interest in botany and natural history, and particularly his work on the new Linnaean system of botanical classification, led to publications in the *Gentleman's Magazine* and the *Philosophical Transactions of the Royal Society*. He was elected a Fellow of the Royal Society in 1762. His book on Linnaeus (also reissued in this series), first published in 1782, was later considered to be of great significance for the acceptance in England of the Linnaean system, and this two-volume work, published in 1790, is still relevant to the study of the history of botany. Volume 1 begins in 'primaeval' and 'druidical' times and continues to the seventeenth century, including the first printed herbals and the work of the great botanist John Ray.

Cambridge University Press has long been a pioneer in the reissuing of out-of-print titles from its own backlist, producing digital reprints of books that are still sought after by scholars and students but could not be reprinted economically using traditional technology. The Cambridge Library Collection extends this activity to a wider range of books which are still of importance to researchers and professionals, either for the source material they contain, or as landmarks in the history of their academic discipline.

Drawing from the world-renowned collections in the Cambridge University Library, and guided by the advice of experts in each subject area, Cambridge University Press is using state-of-the-art scanning machines in its own Printing House to capture the content of each book selected for inclusion. The files are processed to give a consistently clear, crisp image, and the books finished to the high quality standard for which the Press is recognised around the world. The latest print-on-demand technology ensures that the books will remain available indefinitely, and that orders for single or multiple copies can quickly be supplied.

The Cambridge Library Collection will bring back to life books of enduring scholarly value (including out-of-copyright works originally issued by other publishers) across a wide range of disciplines in the humanities and social sciences and in science and technology.

Historical and Biographical Sketches of the Progress of Botany in England

VOLUME 1

RICHARD PULTENEY

CAMBRIDGE UNIVERSITY PRESS

Cambridge, New York, Melbourne, Madrid, Cape Town,
Singapore, São Paolo, Delhi, Tokyo, Mexico City

Published in the United States of America by Cambridge University Press, New York

www.cambridge.org
Information on this title: www.cambridge.org/9781108037327

This edition first published 1790
This digitally printed version 2011

ISBN 978-1-108-03732-7 Paperback

HISTORICAL AND BIOGRAPHICAL

SKETCHES

OF THE PROGRESS OF

BOTANY

IN *ENGLAND*,

FROM

ITS ORIGIN

TO THE

INTRODUCTION OF THE *LINNÆAN* SYSTEM.

BY

RICHARD PULTENEY, *M.D. F.R.S.*

IN TWO VOLUMES.

VOL. I.

LONDON:
PRINTED FOR T. CADELL, IN THE STRAND.
1790.

"Quid quærunt mortales in globo hocce lubrico et horario magis, obtenta supellectili ad vitam maxime necessaria, quam quod levis modo et honesta recordatio nominis—perveniat ad posteros, duretque per aliquot dies ulterius? Quot Heroes, Reges et Imperatores, quot fortes et strenui, non hanc ob causam solam, ingluviem furentis Bellonæ incurrerent, ut modo posteris nomen eorum esset fabula, et cum fabula, memoria? Cur non idem *Botanicis* qui nec minora ausi sunt."

LINNÆUS

TO

SIR JOSEPH BANKS, BART.

President of the Royal Society,
&c. &c. &c.

DEAR SIR,

AS ſoon as I had determined to lay before the public the enſuing Sketches, I could not heſitate in chooſing whoſe name I ſhould wiſh might honour the introduction of them into the world — To whom could a work of this nature, with ſo much propriety be addreſſed, as to him who had not only relinquiſhed, for a ſeries of years, all the allurements that a poliſhed nation could diſplay to opulence and early age, but had expoſed himſelf to numberleſs perils, and the repeated riſk of life itſelf, that he might attain higher degrees of that knowledge, which theſe ſketches are intended to commemorate, in his predeceſſors and countrymen; and as the reſult of which, he has enlarged the ſtock of natural ſcience, beyond all prior example?

That liberality, Sir, with which you impart the fruit of your various labours, and that diſtinguiſhed patronage you ſo amply afford to natural hiſtory at large, and to botanical ſcience in particular, as they demand, ſo have they juſtly ſecured to you, the grateful acknowledgments of all lovers of that ſcience, and of literature, and philoſophy in general.

I have, Sir, on this occaſion only to regret, that my diſtant ſituation has not allowed me, in the compilation of theſe pages, thoſe benefits which your moſt extenſive and valuable library would have held forth to me;

and

and of which, you ſo generouſly permit the communication, to ſuch as deſire to avail themſelves of its advantages.

Permit me then, Sir, to have the honour of inſcribing to you the following SKETCHES, as to an eminent, and no leſs candid judge of the ſubject: and, as a public teſtimony of that moſt perfect reſpect and eſteem, with which I am,

DEAR SIR,

Your much obliged, and

Moſt obedient humble Servant,

RICHARD PULTENEY.

BLANDFORD,
FEB. 28, 1790.

PREFACE.

IN the enlightened ages of Greece and Rome, and under the moſt flouriſhing ſtate of Arabian literature, Botany, as a ſcience, had no exiſtence. Nor was it till ſome time after the revival of learning, that thoſe combinations and diſtinctions were effectually diſcovered, which, in the end, by giving riſe to ſyſtem, have raiſed the ſtudy of plants, to that rank it holds at preſent in the ſcale of knowledge.

If in the contemplation of flowers, mankind at large, have in every age placed one of their pureſt pleaſures, how greatly muſt theſe delights be enhanced to the enamoured votary of Botanical Knowledge! who, whilſt he ſurveys that wonderfully varied elegance and beauty, which charm the eye of all, penetrates ſtill farther, and at the ſame inſtant, diſcerns alſo, thoſe analogies and diſcriminations, in the *number*, *figure*, *ſituation*, and *proportion* of parts, on which are laid the foundations of modern Botanical Science; aſſociations and diſtinctions, which are veiled from the untaught eye of common obſervation, howſoever ſenſible to the general beauties of Nature! And hence, independently of its real and ultimate utility, from the acceſſion of knowledge it brings to the Materia Medica, and by its general aſſiſtance to the various arts and elegancies of life, the ſtudy of the vegetable kingdom, has proved, to numerous ſpeculative and inquiſitive minds, the ſource of much intellectual enjoyment.

This Science is, by many, conſidered as of ſo eaſy attainment, that it is not unuſual to aſſign the name of Botaniſt, to any man whoſe memory enables him to repeat the nomenclature of perhaps a few hundred plants; howſoever uninformed he may be, of thoſe principles which entitle him, to the real name and character: With equal juſtice might any man who knows the names only of the parts of a complex machine, aſſume to himſelf that fame which is due

ſolely

ſolely to the inventor of it. By this degrading idea, men of the firſt learning and talents in this branch of knowledge, have frequently been levelled with the moſt ſuperficial enquirers, and the moſt ignorant pretenders. Hence alſo this Science, which even in a ſpeculative view, holds no mean rank, and, conſidered practically, is cloſely connected with medicine, and with the arts and elegancies of life, has been held forth as a trifling and futile employment. In truth, he properly is entitled, in any degree, to the character of the Botaniſt, whoſe acquirements enable him to inveſtigate, to deſcribe, and ſyſtematically arrange, any plant which comes under his cognizance. But to theſe abilities, in order to compleat the character, ſhould be united, an acquaintance with the Philoſophy of Vegetables, and with the Hiſtory of the Science, in all its ſeveral relations, both literary and practical, from remote antiquity to his own time: attainments which require a competent ſhare of general learning, and no ſmall degree of painful toil and patient induſtry, both in the fields and in the cloſet.

If

If this deſcription of the Botaniſt be a true one, it manifeſtly excludes a number of frivolous pretenders; the ſcience itſelf riſes in importance, and admits of great diverſity of employment, to the taſte, the talents, and learning of thoſe who direct their attention to it. Whilſt then it is the province of ſome to inveſtigate new ſubjects, to aſcertain thoſe imperfectly known, and to record the various improvements and diſcoveries of the day, let it be that of others, to do juſtice to departed merit, to recall the ſcattered remembrances of the lives, and hold out the example of thoſe who have laboured in the ſame field before them.

In tracing the progreſs of human knowledge through its ſeveral gradations of improvement, it is ſcarcely poſſible for an inquiſitive and liberal mind, of congenial taſte, not to feel an ardent wiſh of information relating to thoſe perſons by whom ſuch improvements have ſeverally been given: and hence ariſes that intereſting ſympathy which almoſt inſeparably connects biography with the hiſtory of each reſpective branch of knowledge.

In this age, when succeſsful advancements in the ſtudy of plants, have ſo far extended its pleaſures, as to render Botany almoſt faſhionable; and at a time, when Biographical writings find a reception heretofore unknown; it became matter of ſpeculation, that no one ſhould have delineated the *Riſe and Progreſs of Botany in Britain*, in connexion with the lives of thoſe who have contributed to amplify and embelliſh it.

Among the various enquiries which employ the pens of the learned, none perhaps afford more general ſatisfaction, than ſuch as relate to the origin and progreſs of ſcience and literature. But when theſe lead to objects which we love and cheriſh, they come recommended by a charm that ſecures a welcome, and thus promiſe a more peculiar entertainment and gratification: however, diſquiſitions of this kind are of difficult execution, eſpecially when applied to ſubjects of a ſcientific nature, as requiring the union of various talents in the writer—an appropriate ſhare of learning, an extenſive literary as well as practical acquaintance with the ſubject, united to all thoſe qualifications re-

quiſite

quiſite in a biographer, ſuch as diligence and accuracy in inveſtigating the diſcoveries of his authors, and impartiality in characterizing them, and in aſſigning to each his due degree of merit. To theſe perſonal requiſites muſt be added, the adventitious circumſtances of a ſituation favourable to his reſearches, not only from manuſcripts, and large libraries, but from actual intercourſe with the learned.

Fully ſenſible in this view of the little claim I have to the character and advantages here ſpoken of, it becomes neceſſary, to avoid the cenſure of temerity, that I ſhould premiſe ſome account of the original occaſion of this attempt.

The attention I had given to Engliſh Botany in my younger days, had prompted me, at one time, to plan a *Flora* of the plants of this kingdom, on an extenſive ſcale; including, beſides the medical and œconomical hiſtory of each, a *Pinax*, in which it was my deſign to have diſtinguiſhed, as far as I was able, the firſt diſcoverer of each ſpecies, both among foreign writers and thoſe of our own kingdom; and to have

arranged

arranged all their ſynonyms, at large, under each plant, in chronological order. To ſuch a work the following *ſketches*, in a ſomewhat more contracted form, were intended as an introduction. In the mean time, if more important avocations had not, the want of neceſſary aſſiſtance from books, would probably have ſtopped the progreſs of a plan of ſuch extent. Although this purpoſe was relinquiſhed, yet, as the materials were collected, and this part of the deſign was independent of the other, I flattered myſelf, that, having made ſome alterations, and enlarged the whole, under ſo total a want of any ſimilar work, theſe anecdotes might afford information to young Botaniſts, and poſſibly ſome amuſement to thoſe of more advanced knowledge in the ſcience.

Although botanical writings are the principal objects of theſe pages, yet, as ſeveral of theſe authors were conſpicuous for their various attainments in different branches of literature, their other purſuits and publications, where my reſources have afforded opportunity, have occaſionally been recited;

and

and I have been more particularly ſolicitous to collect into one view, under each author, thoſe various temporary and occaſional productions, which, after the eſtabliſhment of the Royal Society, were communicated to that body, and form a part of the *Philoſophical Tranſactions*.

In conſidering the botanical writings, eſpecially thoſe of the firſt eminence, I have had recourſe, with few exceptions, to the books themſelves; but, confined to a private collection, have yet too frequently had occaſion to regret the want of more extenſive aſſiſtance; and, although I have not formally quoted my authorities, on every occaſion, they will be ſufficiently manifeſt to all ſuch as are converſant in botanical literature. In the hiſtorical and biographical parts, the moſt material and authentic facts, have likewiſe been derived from the reſpective authors in botany: and, not unfrequently, I have availed myſelf of ſeveral of the older periodical publications. Excluſively of theſe, beſides collateral aſſiſtance received from ſeparate works, and from various collections of ſmaller bulk, I more eſpecially acknowledge my obligation to the authors recited below *.

In a work intended to exhibit the progreſs of the ſcience in *England*, and to aſſign to each writer his reſpective praiſe, I could have wiſhed to have ſubjoined a com-

* GESNERI, Bibliotheca Univerſalis. fol. *Tigur.* 1545. et *ejuſd.* Epitome à *Simlero* et *Friſio.* fol. 1583. item, *eſuſdem* Præfatio in Libros de Natura Stirpium *H. Tragi.* 4°. *Argent.* 1552.

PHILOSOPHICAL TRANSACTIONS, 4°.

Van der LINDEN, De Scriptis Medicis à *Mercklino.* 4°. *Norimb.* 1686.

HERBELOT, Bibliotheque Orientale. fol. 1697. *Maeſtricht.* & 1776.

WOOD, Athenæ Oxonienſes. fol. *Lond.* 2 vol. 1721.

TOURNEFORT, Iſagoge in Rem Herbariam. in Rei Herbariæ Inſtitutionibus. 4°. *Paris,* 1719.

BOERHAAVE Methodus Studii Medici. 8°. 1710. Emaculata et aucta ab Hallero. 4°. 2 vol. *Amſt.* 1751.

CONRINGII, Introductio in Univerſam artem Medicam. 4°. 1726. *Hal.*

FRIEND, Hiſtory of Phyſic. 2 vol. 8°. 1727.

LE CLERC, Hiſtoire de la Medicine. 4°. *à la Haye.* 1729.

MANGETI, Bibliotheca Scriptorum Medicorum. 4 vol. fol. *Gen.* 1731.

GENERAL DICTIONARY, 10 vol. fol. 1734—1741.

LINNÆI, Bibliotheca Botanica. 8°. 1737. *Amſt.* et 1751.

SEGUIER, Bibliotheca Botanica. 4°. *Hagæ Com.* 1740.

TANNER, Bibliotheca Britannico-Hibernica. fol. *Lond.* 1744.

SCHMIEDEL, in Præfat. ad Geſneri Opera. fol. *Norimb.* 1753.

BIOGRAPHIA BRITANNICA. fol. *Lond.* 7 vol. & 2d edit. 4 vol.

MATTHIAS, Conſpectus Hiſtoriæ Medicorum. 8°. *Gotting.* 1761.

FABRICII, Bibliotheca Latina, 2 tom. 4°. 1723 & 1734, & 3 tom. 8°. ab *Erneſto Lipſ.* 1773.—*ejuſdem* Bibliotheca Latina mediæ et infimæ latinitatis, 6 vol. 8°. *Hamb.* 1735—1746.

HALLER, Bibliotheca Botanica. 2 tom. 4°. 1772.

GRANGER, Biographical Hiſtory of England, 4 vol. 8°. 3d edit. 1779.

NOUVEAU DICTIONNAIRE HISTORIQUE. 8° 1765. 6ieme edit. 8 tom. *Caen.* 1786.

ELOY, Dictionnaire Hiſtorique de la Medicine. 4°. 4 tom. *Mons.* tom. 1778.

plete

plete catalogue of all the Engliſh plants, with the names of the firſt diſcoverer annexed; or of that author in whoſe work each firſt occurs, as an Engliſh ſpecies. The progreſs I had made in the intended Pinax above-mentioned, would have enabled me to have made this addition; but, as ſuch a catalogue could have afforded gratification only to the more curious and critical botaniſts, unleſs thrown into a form, by the addition of other matter, which would have increaſed the bulk of this work to another volume, it was judged moſt proper to omit it.

Conſcious of the many defects attending theſe *ſketches*, and fully ſenſible that they merit no higher appellation than what the title imports, it is with much deference, even under that idea, that I ſubmit them to the inſpection of the literary world; and, perhaps, the indulgence they require, is greater than ought to be expected: but I am willing to hope, that they will find that reception from learned and candid judges, which ſuch are wont to beſtow on a firſt eſſay, in any department of literature.

TABLE of CHAPTERS

IN VOLUME I.

TABLE OF CHAPTERS.

TABLE OF CHAPTERS.

VOL. I.

Errors in the Printing.

Page 249. line 8. *For* CAMBDEN, *read* CAMDEN.
256. — 16. — apophthegms — apothegms.

HISTORICAL AND BIOGRAPHICAL SKETCHES OF THE PROGRESS OF BOTANY, IN ENGLAND.

CHAP. I.

The origin of Botany in general—Its ſtate in the druidical times—Rites obſerved by the Druids *in collecting the* miſſeltoe, vervain, *and* ſelago—*All but the* miſſeltoe *difficult to be aſcertained—Of the* herba Britannica, *and the* roan-tree.
Saxon *Botany—Manuſcripts extant in that language—*Saxon *verſion of* Apuleius.

PRIMÆVAL BOTANY.

THE origin of Botany, conſidered in the moſt extenſive view, muſt have been coeval with man. Before the invention of arts, the diſcovery of metals, and the uſe of implements and arms, by which animals were more immediately ſubjected to their power, it muſt be ſuppoſed that the human race derived, from the vegetable creation, the chief part of their ſuſtenance, and the primary conveniences of life. Roots, fruits, and herbs, muſt then have conſtituted the food of man. Trials, and experience, would teach him all that choice and variety, which his different ſituations allowed. The ſame faithful directors would inſenſibly inform him of the various qualities, and the different effects of them on his body. As the ſphere of his obſervations and experience enlarged, he would derive the knowledge, and diſtinction, of ſuch as were of eaſy, or of difficult digeſtion. He would diſcover the flatulent kinds, and ſuch as corrected flatulency: which opened, or which conſtipated, the body; which was moſt nutritive, and probably, by fatal accidents, which were poiſonous. Hence the rudiments of medical ſcience.

This various knowledge would be handed down traditionally, from one generation to another, and with it, the names of ſuch as were happily the firſt diſcoverers of new aliments, or medicinal properties, would descend with increasing reverence, until, involved in obscurity by length of time, superstition raised them to the rank of gods. Thus, in the early ages of mankind, as now among the still unlettered and uncultivated nations of the earth, the administration of simples, for the cure of wounds and diseases, was almost ever accompanied with superstitious ceremonies and incantations. Hence too, in process of time, the character of the priest and the physician was united; and the sick resorted to the temples of the gods for relief: and, although investigation and rational science made slow progress, yet, in every nation, from the most cultivated to the most barbarous, the number of simples used for medicinal purposes, became by degrees very considerable. Thus, when at length, physic assumed a more regular form, and was taught in the schools of Greece, the writings of HIPPOCRATES enumerate three hundred vegetables used in physic. Four centuries afterwards they were augmented by DIOSCORIDES to near seven hundred; and to these the Arabians added no inconsiderable number of valuable articles. There is room to believe, that the antient *Gymnoſophiſts* of the Eaſt, purſued the ſtudy of plants, with a ſucceſs equal to that of the Greeks; and the modern nations of the Eaſt, the Japoneſe, the Chineſe, and the Brachmans of India, inconteſtibly excel the enlightened nations of Greece and Rome, in their knowledge of Botany: witneſs the " Garden of Malabar," which comprehends near eight hundred plants; all which are deſcribed, and the virtues recorded, with an accuracy and preciſion, unexampled in the antient authors of *Greece* and *Rome*. But to approach nearer home: the Druids of *Gaul*, and of *Britain*, cultivated the knowledge of herbs, with no inconſiderable diligence. Whether theſe antient *Magi* of the Weſt, who were both prieſts and phyſicians, ſprung from thoſe of the Eaſt, and thus derived their knowledge from a common ſource, a point which has hitherto divided the learned, or, whether their ſcience was the reſult of their own inveſtigation, I muſt leave to the critical antiquary to determine.

DRUIDICAL BOTANY.

In the mean time, in tracing the origin and progreſs of botanical ſcience in *Britain*, a ſurvey of its ſtate in the druidical times, ought to claim the firſt attention; but in fact, the little information tranſmitted to us from the antients, relating to this extraordinary ſect, being almoſt wholly confined to *Cæſar* and *Pliny*, precludes any enlarged view reſpecting my particular object. It is from *Pliny* we learn, that to the *miſſeltoe*, the *vervain*, the *ſelago*, and the *ſamolus*, theſe antient fathers of druidiſm attributed efficacies almoſt divine; and ordained the collection, and adminiſtration of them, with rites and ceremonies, not ſhort of religious ſtrictneſs, and ſuch as countenanced the groſſeſt ſuperſtition.

The *miſſeltoe*, for inſtance, muſt be cut only with a golden knife; muſt be gathered when the moon was ſix days old; the prieſt cloathed in white; the plant received on a white napkin; and laſtly, two white bulls were to be ſacrificed; and thus con-

 ſecrated,

ſecrated, miſſeltoe was an antidote to poiſon and prevented ſterility *.

The *miſſeltoe* perhaps, is, of theſe plants, the only one fully aſcertained at this time. Its paraſitical growth, the preference which the Druids gave to that which grew on the oak, aſſiſted by the deſcriptions the antients have left of it, will ſufficiently juſtify the application to the *viſcum* of the moderns. May I not add, that probably, amidſt the manifold virtues antiently aſcribed to this plant, its power of curing the falling-ſickneſs, which has accompanied it almoſt to the preſent time, is the remnant of druidical uſe and tradition?

The *vervain*, after previous libations of honey, was to be gathered at the riſing of the dog-ſtar; when neither ſun nor moon ſhone; with the left hand only; after deſcribing a circle round the plant, &c.; and thus prepared, it vanquiſhed fevers, and other diſtempers; was an antidote to the bite of ſerpents, and a charm to conciliate friendſhip †.

* *Pliny*, lib. xvi. c. 44.

† *Ib.* lib. xxv. c. 9.

With

With reſpect to this herb, the *hierobotane*, the *ſacra herba* of DIOSCORIDES, although the modern botaniſts have now agreed to confine the term to the *verbena*, which PLINY has deſcribed, as having narrower and ſmaller leaves than the oak, it may be remarked, that there has been a diverſity of opinions among the commentators, relating to the plant; and it is acknowledged that *verbena* or *verbenacea*, was alſo applied, as a general term for all plants uſed about the altar in ſacrifices. To this day the Tuſcans apply the word *vervena* to ſlips, ſhoots, ſuckers, or bundles of plants of any kind.

The *ſelago* was not to be cut with iron; nor touched with the naked hand, but with the *ſagum*; the Druid cloathed in white, and his feet naked, with other magic ceremonies. Thus collected, and conſecrated, it became a remedy for diſeaſed eyes, and a charm againſt misfortunes*.

It is, nevertheleſs, equally difficult to determine the *ſelago* of the Druids; PLINY

* *Ib.* lib. xxiv. c. 11.

having only deſcribed it as like the *ſavin*; a deſcription which will accord with a variety of plants of Europe. Moſt authors, nevertheleſs, have agreed, from this reſemblance, to conſider it as a ſpecies of *wolfs-claw moſs*, which is now called *lycopodium ſelago*. Cæsalpinus, however, thinks it was a *ſedum*; and Guilandinus, an *erica*, or heath, and probably with more reaſon.

Various, but equally ſuperſtitious, were the rites attendant on the *ſamolus*, which was given to preſerve oxen and ſwine from diſeaſes.

This is a plant of which ſtill greater doubts remain, Pliny having ſaid nothing further of it, than that it grew in moiſt places. Hence the name is applied to a plant called *round-leaved brooklime*; but, as forming a ſeparate genus in modern arrangements, it has acquired the name given as above, from *Pliny*. Others have thought it a ſpecies of *pulſatilla*, or paſque-flower; ſince one of that kind retains, among the Bologneſe, the name of *ſamiglo*.

The ſame uncertainty attends all diſquiſitions relating to the *herba Britannica*, of

Dioscorides

DIOSCORIDES and PLINY, famed for having cured the soldiers of *Julius Cæsar*, on the Rhine, of the *Scelotyrbe*, or the disease supposed to be our sea scurvy. The uses of this herb were thought to have been derived from the Britons; the name suggested this notion; but later etymologists have found a different derivation: i. e. Brit. *consolidare*; Tan. *Deus*, *Ica* s. Hica, *ejectio*; unde, *Britannica dicitur herba, quæ firmet et consolidat dentes vacillantes* *. The commentators have applied the description given by those two antients, to a variety of simples. By some, it has been thought to be the *polygonum persicaria*, or spotted arsmart: by others, the *primula auricula*, or wild auricula: by our own first herbalist, TURNER, who observed it plentifully in *Friesland*, the scene of *Pliny*'s observations on its effects, the *polygonum bistorta*, or bistort: at length, *Abraham* MUNTING, a Dutch physician, published a treatise in 1681, professedly to prove, that the *Britannica* was the *hydrolapathum magnum*, (*ru-*

* RAY. *Hist. Plant.* i. p. 172.

mex

mex aquaticus) or great water dock. In this opinion RAY, and others, have acquieſced.

I ſhould not have dwelt ſo long on theſe circumſtances, but to ſhew the mortifying uncertainty attending the application of the names of plants from the antients, ariſing from their vague and indeciſive deſcriptions. I add, that Mr. LIGHTFOOT thinks, there are ſufficient traces in the highlands, of the high eſteem in which the Druids held the quicken-tree, or mountain aſh; *ſorbus aucuparia*. It is, more frequently than any other tree, found planted in the neighbourhood of druidical circles of ſtones, ſo often ſeen in Scotland. Poſſibly this fact may be more equivocal than the ſuperſtitious uſes to which it is ſtill applied. It is believed, that a ſmall part of this tree carried about them, is a charm againſt witchcraft and enchantment. The dairy-maid drives the cattle with a ſwitch of the *roan-tree*, for ſo it is called in the highlands, as a ſecurity againſt the ſame direful evils; and in one part of *Scotland*, the ſheep and lambs are, on the firſt of May, ever made to paſs through a hoop of *roan-wood*.

Short,

Short, and imperfect, as this view of Druidical Botany may be, as delivered to us by PLINY, yet there can be no doubt that the Britons, like all other rude nations, drew their medicinal ſources from the ſimples growing around them, and were therefore well acquainted with common plants. And, although there are not, as far as I know, any herbals extant in the antient Britiſh language, or in any tranſlation from it, by which the degree and extent of their knowledge may be preciſely aſcertained; yet, as far as reſpects the nomenclature merely, ſome reaſonable eſtimate may, I apprehend, be formed from the liſt of *Welch* names of plants, preſerved by GERARD, as communicated to him by Mr. *Davies* of *Guiſſaney*, in *Flintſhire:* from the *Iriſh* names, as we find them in Mr. HEATON's catalogue, printed in THRELKELD's *Synopſis;* to which I may add, the *Erſe* names communicated by the Rev. Mr. *Stuart*, to the late excellent and much-lamented botaniſt, the Rev. Mr. LIGHTFOOT. Theſe liſts might, without doubt, be greatly amplified, by the aſiduity of ſkilful botaniſts well

well versed in the respective languages. THRELKELD's list, which is the most copious, comprehends near four hundred names; and the analogy perceivable between these and the *Erse* names, sufficiently marks a common origin. I am tempted to produce a few instances *.

SAXON

* Muiriunagh. *Irish.* Muran. *Erse.*	*Arundo arenaria s. Spartum.* Sea Matweed.
Cruah Phadruig. *Irish.* Cuah Phadruic. *Erse.*	*Plantago Major.* Great Plantain.
Slan luss. *Irish.* Slan lus. *Erse.*	*Plantago lanceolata.* Ribwort Plantain.
Cran Tromain. *Irish.* An druman. *Erse.*	*Sambucus.* Elder-tree.
Fraogh. *Irish.* Fraoch. *Erse.*	*Erica.* Heath.
Feirdriss. *Irish.* An-Fhearr-driss. *Erse.*	*Rosa canina.* Dog Rose.
Carmel. *Irish.* Cor, Cormeille. *Erse.*	*Orobus sylvaticus.* Wood-Pease.
Tæd Coluim Kille. *Irish.* Acklasan-Challum-chille. *Erse.*	*Hypericum perforatum.* St. John's Wort.
Meacan tovach. *Irish.* Mac-an-dogha. *Erse.*	*Arctium Lappa.* Bur-dock.
Liagh Luss. *Irish.* An-liath lus. *Erse.*	*Artemisia vulgaris.* Mugwort.
Gallan. *Irish.* An-gallan-mor. *Erse.*	*Tussilago Petasites.* Butter-Bur.
Noinin, nonin. *Irish.* Noinein. *Erse.*	*Bellis perennis.* Daisy.

Ahair

SAXON BOTANY.

The hiſtory of *Saxon* Botany muſt be very ſhort. No nations, however rude, have yet been diſcovered, who were ſo regardleſs of health, as not to have a knowledge of, and ſome dependence upon, the virtues of certain ſimples. There is ſufficient evidence, that our Saxon anceſtors did

Ahair Talham. *Iriſh.* A'chaithir-thalmhain. *Erſe.*	*Achillæa Millefolium.* Yarrow, or Milfoil.
Sail Tovagh. *Iriſh.* Sail Chuach. *Erſe.*	*Viola odorata.* Sweet Violet.
Beihe. *Iriſh.* Am-Beatha. *Erſe.*	*Betula alba.* Birch-Tree.
Fearnog. *Iriſh.* Am-Fearna. *Erſe.*	*Betula Alnus.* Alder-Tree.
Cran Darrah. *Iriſh.* An Darach. *Erſe.*	*Quercus Robur.* The Oak.
Guiſagh. *Iriſh.* An Guithas. *Erſe.*	*Pinus ſylveſtris.* Wild Pine. Scotch Fir.
Soileog. Saileagh. *Iriſh.* Sileach. *Erſe.*	*Salix alba.* Willow.
Ruideog. Raodagh. *Iriſh.* Roid. *Erſe.*	*Myrica Gale* ; or, Sweet Myrtle.
Beecora lecra. *Iriſh.* Beeora leacra. *Erſe.*	*Juniperus.* *Juniper.*
Raineagh muire. *Iriſh.* Raineach. *Erſe.*	*Pteris aquilina.* Fern, or Brakes.
Garvogagh. *Iriſh.* Garbhag-an-t-ſleibh. *Erſe.*	*Lycopodium Selago.* Wolfs-claw Moſs.
Duilleaſg. *Iriſh.* Duilleoſg. *Erſe.*	*Fucus palmatus.* Sweet Fucus. Dulſe.

not

not wholly disregard this study; since, although rare, there are manuscript Saxon herbals extant in several public libraries. The two following occur in the Bodleian:

4123. HERBARIUM. Saxonice.

5169. LIBER MEDICINALIS, *continens virtutes herbarum.* Saxonice.

I am unable to determine whether the above are the same with the two following, which Dr. *Ducarel* notices from the Harleian collection:

5066. Entitled, HERBARIUM. Saxonice.

585. *Tractatus, qui ab Anglo-Saxonibus dicebatur* LIBER MEDICINALIS.

The last is said to be an Anglo-Saxon version of APULEIUS, whom I shall have occasion to mention hereafter. The date of this translation is of the tenth century. The Saxons having been converted to Christianity at the latter end of the sixth century, the communication between *Britain* and *Rome* became by degrees very frequent,

quent, and learning was then firſt introduced into theſe realms.

The golden age, if I may be allowed that expreſſion, of the Anglo-Saxon learning, was the reign of ALFRED the Great. That munificent prince not only himſelf tranſlated Latin authors, but, as hiſtorians inform us, encouraged in every way, the tranſfuſion of all the knowledge of the times into the common language of the kingdom. To this æra, therefore, may reaſonably be referred the Saxon verſion of APULEIUS; whoſe book ſeems to have preſerved popularity through all the middle ages, and was found in common uſe at the æra of printing.

As no publication of any *Saxon* herbal has ever taken place, we are unable to define the extent of the knowledge of that time: at preſent, therefore, as in the inſtance of antient *Britiſh* Botany, we can only recur to the nomenclature of the indigenous names, by which ſome of them are yet known; although many others have given way to Greek and Latin terms, and ſome to other revolutions, occaſioned by the gradual

gradual progreſs of reformation throughout the ſcience in general.

A liſt of the *Anglo-Saxon* names would be recoverable, in a great degree, by recurring to the old herbals, to SKINNER's Lexicon, and other authorities of that kind. It would, I am perſuaded, be more extenſive than a ſuperficial view might ſuggeſt, and would do credit to our Saxon anceſtors I cannot help remarking, that many miſtakes have probably ariſen from the neglect of our firſt reformers of Botany in *England*, after they had formed ſcientific names, in not preſerving alſo the old and provincial terms; and that, on the whole, this neglect has retarded the progreſs of knowledge on this ſubject.

CHAP.

CHAP. 2.

General ſtate of Botanical knowledge during the dominion of the Saracens—*Corrupt tranſlations of* Dioſcorides—Avicenna—Aſchard, *or* Ebn Beithar, *the capital Writer in Botany among the* Arabians—Schola Salernitana—Engliſh *Writers during the middle ages*—Henry *of* Huntingdon — Arviel — Bray — Legle, *or* Gilbertus Anglicus—Ardern—Daniel—Bollar—Horman—*MSS. of anonymous Authors—Tranſlations and editions of* Apuleius *and* Macer, *in uſe in* England *at the invention of printing—Specimen of the ſuperſtition of* Apuleius.

MIDDLE AGES.

LEARNING and ſcience follow the fate of empires. On the decline of thoſe of *Greece* and *Rome*, and during that period in which the Saxons were eſtabliſhing themſelves in *Britain*, medical knowledge paſſed into the hands of the triumphant Saracens. *Bagdat*, under the Eaſtern Caliphs, became the ſeat of learning. Much of the Greek phyſic and philoſophy was corruptly tranſlated by the command of Muſſelmen; among whom at length it received due reception and encouragement. Schools were eſtabliſhed, in which ARISTOTLE, GALEN, DIOSCORIDES, and other writers, were ſtudied; and their doctrines at length pervaded the whole dominion of the Saracens, and finally flouriſhed in the univerſities of *Spain*.

DIOSCORIDES, though in a corrupt and mutilated ſtate, formed the baſis of knowledge in the *Botany* and Materia Medica of the Arabians. The ſituation of *Bagdat*, and its connection with *India*, allowed them ſcope to introduce into phyſic ſeveral uſeful ſimples. Among others, we owe to theſe Orientals the milder purges of the preſent day; ſuch as *ſenna*, *caſſia fiſtula*, *manna*, *tamarinds*, *rhubarb*, and ſeveral drugs of other qualities, of which ſome retain a place in the preſent reformed ſtate of the *Materia Medica*. AVICENNA, we are told, had coloured drawings for the inſtruction of his pupils in Botany; and *Proſper* ALPINUS aſſures us, he ſaw at *Cairo* a volume of paintings of the plants of *Ægypt*, *Arabia*, and *Ethiopia*, which had been done for the uſe of a Sultan.

It

It is not eaſy, however, to judge, with preciſion, of the extent of Arabian knowledge on the ſubject of our work; ſince, probably, the beſt book of the Arabian ſchool has yet remained unpubliſhed, that of *Ebn* BEITHAR. It is extant in the Pariſian, the Eſcurial, and other libraries. This learned Arab was particularly attached to the botanical branch of phyſic. He was born in *Spain*; and after viſiting *Africa*, travelled into the Levant, *Aſia*, and even as far as the Indies, to improve his knowledge. In his return he was patroniſed by *Saladin*, at *Cairo*, and died in 1248.

HERBELOT informs us, that from the ſuperiority of his learning in this branch, he was ſtyled *Aſchard*, or The Botaniſt. He wrote "A General Hiſtory of Simples, or of Plants, ranged in alphabetical order;" in which he gives the Greek, Arabic, and vernacular names; with the deſcriptions of each; and particularly, in a more detailed manner, thoſe not deſcribed by DIOSCORIDES and PLINY.

There is, notwithſtanding, but little room to believe, that more original knowledge could be derived from the Arabian monuments of ſcience in this, than in the other departments of phyſic. In their beſt authors, even the Greek names of plants are ſo groſsly perverted, that they are ſcarcely to be known.

The Grecian authors having been inaccurately tranſlated at firſt, and the language neglected afterwards, phyſic loſt much under the dominion of the Arabians. It was, in the end, a corrupt Galenic theory, with an admixture of *aſtrology* and ſuperſtition. In this ſtate the learned of Europe found it, in the celebrated Mooriſh univerſities of *Spain*. In the weſtern parts of Chriſtendom, eſpecially after the lapſe of the Latin tongue in *Italy*, it was ſcarcely leſs obſcured by the ignorance of the Monks, by whom, almoſt ſolely, the practice of it was engroſſed.

Even the firſt univerſity in Chriſtendom, the renowned ſchool of *Salernum*, founded by *Charlemagne* in the beginning of the ninth century, received its dictates from the corrupt ſources of the Arabians; whoſe works are ſaid to have been at length tranſlated into

into Latin by *Constantine* the African. The famous precepts *de Conservanda Valetudine*, issued from that school for the use of *Robert* duke of Normandy, were, without doubt, well known in England, and probably excited attention to the study of Vegetables; concerning which, numerous rules and cautions occur in that remnant of the learning of those days.

During all these ages, the original sources in the Greek authors were almost wholly forgotten, and the productions of that long night of science were equally rare and unimproving.

I shall, nevertheless, enumerate briefly a few of those English authors, who were most conspicuous for any attention to the simples used in medicine, which alone bounded the botanical knowledge of those times.

One of our earliest writers, after the Conquest, was the historian HENRY *Archdeacon of Huntingdon*, in the time of king Stephen and Henry the IId. *Bishop* TANNER informs us, that he left a MS. in eight Books, *De* HERBIS, *de Aromatibus, et de Gemmis*. Bib. Bodley. 6353.

Of nearly the ſame age are ſaid to be, ſome manuſcripts preſerved in *Bibl. Regia Lond.* under the following titles, *De Natura Pecudum*, ARBORUM, *et Lapidum:* and one De *Naturis Herbarum*. Biſhop *Tanner* mentions an Engliſhman of the name of *Henry* ARVIEL, who had travelled much, and reſided ſome time at *Bologna*, about the year 1280. He left a manuſcript *De Botanica, ſive Stirpium Varia Hiſtoria*.

The ſame author notices a manuſcript, in the Sloanean collection, of *John* BRAY, who lived in the time of Richard the IId. He ſtudied Botany and Phyſic, and received an annual penſion from the king, for his knowledge and ſkill in theſe ſciences. It is entitled, *Synonyma de nominibus Herbarum*. It contains the names, in Latin, French, and Engliſh.

Beſides the *Compendium Medicinæ* of GILBERTUS LEGLE, or GILBERTUS ANGLICUS, who alſo flouriſhed in the thirteenth century, a manuſcript is recorded of that author, under the title of *De re Herbaria, lib.* 1. and others, *De Viribus et Medicinis Herbarum, Arborum et Specierum: et de Virtutibus Herbarum, lib.* 1.

The famous Engliſh ſurgeon *John* ARDERN of Newark, extolled by Dr. *Friend*, as the reviver of ſurgery in England, who flouriſhed ſoon after *John* of GADDESDEN, in the middle of the fourteenth century, left a manuſcript, which is in the Sloanean library, under the title of *De re Herbaria, Phyſica, et Chirurgica.*

Henry DANIEL, a Dominican friar, ſaid to be well ſkilled in the natural philoſophy and phyſic of his time, left a manuſcript inſcribed *Aaron Danielis.* He therein treats *De re Herbaria, de Arboribus, Fructicibus,* &c. He flouriſhed about the year 1379.

Appertaining to my ſubject I alſo mention, a treatiſe, written, as is ſuppoſed, in the time of Edward the IIId. by WALTER *de* HENLEY, entitled, *De Yconomia ſive Houſbrandia*; in which, Biſhop *Tanner* ſays, he has treated his ſubject well, according to the uſage of the time.

NICOLAS BOLLAR, educated at *Oxford*, whom

whom TANNER repreſents as eminent for his knowledge in natural philoſophy, wrote *De Arborum Plantatione*, lib. 3. *De Generatione Arborum et modo Generandi et Plantandi*, lib. 2. and other tracts now in manuſcript.

There is a manuſcript ſaid to be preſerved in Baliol college, written by JOHANNES *de* S. PAULO, *De Virtutibus Simplicium Medicinarum*. The age of theſe two laſt is not ſufficiently aſcertained; neither is that of a manuſcript in *Caius* and *Gonville* college, *Cant.* entitled *Cinomia* (Synonymia) *Herbarum*.

The following authors, who wrote, at leaſt prior to the introduction of printing into England, are enumerated, by Biſhop *Tanner*, and others.

Henricus CALCOENSIS, a prior of the Benedictine order, is ſaid, by *Dempſter*, to have travelled into *France*, *Germany*, and *Italy*, ſolely to enjoy the converſation of the learned. He wrote *Synopſis Herbaria*, Lib. 1. and tranſlated PALLADIUS *de re Ruſtica*, into the Scottiſh tongue, about the year 1493.

William

William HORMAN, a native of *Saliſbury*, was educated at Wincheſter ſchool, and became a perpetual ſellow of New College in 1477. In 1485 he was choſen ſchoolmaſter and fellow of Eton, and at lengt elected vice-provoſt of the ſame college. He was a man of extenſive and various erudition. Among numerous productions, he left a book under the title of *Herbarum Synonyma*. He wrote indexes to the antient authors *De re Ruſtica:* to *Cato*, *Varro*, *Columella*, and *Palladius*. After ſeveral years of retirement, he died in 1535, and was buried in the chapel of the college.

The writers, and the age, of the two following manuſcripts, are unknown.

Liber de Herbis, in the library of Corpus Chriſti.

Nomenclatura Vocabulorum in Medicina receptorum, præſertim etiam Herbarum; in the library of Magdalen college.

The underwritten, without any author's names, are in the Aſhmolean library, with the annexed dates.

Diverſe phyſical receipts with an Herbal, 1438, N° 7704.

An

An HERBAL, Alphabeticum, 1443, N° 7709.

An HERBAL, in old Englifh, 1447, N° 7713.

Phyfical Plants, Englifh, 1481, N° 7724. Alfo,

A *defcription of fome fimples*—In the Bodleian library, N° 2073.

Exlufive of many others, more ftrictly medical, the under-written * anonymous manufcripts, though the dates have not been precifely determined, are, with good reafon, fuppofed to have been written, if not prior to the invention of printing, at leaft before the introduction of that art into England.

This

* In the Bodleian library.

2543. *Anonymus, de Arboribus, Aromatis, et Floribus.*
2062. An Herbal.
2562. *Gloffarium Latino-anglicum Arborum, Fructuum, Frugum, &c.*
2335. *Nomina Herbarum, Latine, Gallice, Anglice.*
2257. Concerning the Virtue of fome Herbs.
2072. *De fedecim Herbis et earum Virtutibus.*
1798. *Herbarium.*

3828.

This liſt, perhaps already too long, might have been conſiderably extended, but that it would have unneceſſarily ſwelled this article. As none of theſe manuſcripts, however, have been publiſhed, the exact ſtate and progreſs of the ſcience cannot be aſcertained; yet enough is ſeen to convince us, that, although its advancement was ſlow and inconſiderable, it was not wholly loſt in the darkneſs of that night, which, for ſo many ages, obſcured the ſources of knowledge. It is highly probable, that very

3828. *Herbarium Anglico-latinum alphabeticum.*
6206. *De-Plantis admirandis.*
2073. Deſcription of ſome Simples.
2626. *Lexicon Medicamentorum Simplicium.*

In the Aſhmolean library.

7762. *Alphabeta de diverſis Nominibus Herbarum.*
7541. *De Naturis quarundam* (animatium) *Arborum,* &c. cum Iconibus pictis.
7778. *Catalogus Plantarum, additis,* ſubinde, Nominibus Anglicis.
1397. *De Dicta Salutis, et Catalogus Plantarum.* Lat. Angl.
7634. "An alphabetical Catalogue of Plants."
7537. "A Book of Plants, delineated in their natural Colours."
7694. "Alphabetical Catalogue of Plants."

In

very few of the manuſcripts before enumerated, exhibit any conſiderable portion of original matter; but, that they are principally extracts and compilations, from preceding writers of the lower age; ſuch as, *Apuleius*, *Æmilius Macer*, *S. Sethus*, *Iſidore*, *Conſtantinus*, the *Pandects* of *Matthew Sylvaticus*, *Platearius*, ſome of the later Arabians, and other writers of that ſtamp. At the renovation of knowledge juſt mentioned, theſe appear to have been the primary ſources from which our anceſtors of that generation derived aſſiſtance; ſince we find many MSS. of the above au-

In other collections the following:

976. *Tractatus de Herbis.* Bibl. Caj. Gonv. Cant.
8875. "The Book of Simples; or a Treatiſe of Herbs and their Virtues." Sloan.
1747. *De Herbis et Plantis.* Coll. John. Bapt. Oxon.
1695. *Notabilia de Vegetabilibus et Plantis.* Bib. S. Petri Cant.
844. *Nomina Herbarum, earumque Vires.* Bib. Caj. Gonv.
8738. *Nomina Herbarum, et de earum Proprietatibus.* Sloan. *an? idem cum priori.*
959. *Alphabetum Herbarum, cum Synonymis.* Bib. Caj. Gonv.
8746. *Des Proprietés et Noms des Herbes.*

thors

thors were in being, at the origin of printing, and were early iſſued from the preſs as the manuals of that day, in various parts of Chriſtendom.

It has been obſerved, that the laſt-mentioned Saxon manuſcript, was a tranſlation of *Lucius* APULEIUS *Madaurenſis*; whoſe work, from ſeveral other circumſtances, there is room to believe, was, at that time, more diffuſed and popular in *England*, than any other. This author, who lived in the age of the *Antonines*, was born at *Madura* in *Africa*, at that time a ſeat of learning. He afterwards ſtudied at *Carthage*, and at *Athens*, and for ſome time applied himſelf at *Rome* to juriſprudence, but at length quitted it, and devoted himſelf wholly to philoſophy and phyſic. He is well known as the author of the *Mileſian* Fables, and other works of learning. His book *De Herbis, ſive de Nominibus ac Virtutibus Herbarum*, alone comes under our cognizance: In this he recites the names of medicinal herbs, in the Greek, Latin, Egyptian, Punic, Celtic, and Dacian, and of ſome in the oriental languages.

These

Theſe names form the bulk of the book, which conſiſts of one hundred and thirty chapters. After each name follows a ſhort deſcription of the plant, the place of growth, and the properties. Then the diſeaſes to which each ſimple is applicable. The work nevertheleſs abounds with groſs errors in the names of plants, and inculcates the moſt abſurd ceremonies and ſuperſtitions in the adminiſtration of remedies; yet it was in much eſteem throughout the dark ages of literature.

It muſt not however be concealed, that ſome of the learned have judged, that this work, at leaſt as it now appears, was not written by the author whoſe name it bears, but at a much later period. JOHNSON, the editor of *Gerard*, imagined it to be a tranſlation of a Greek writer of the eighth century; but his conjecture is not thought probable by *Fabricius* *. The remarks of *Johnſon* prove, that this work was in common uſe in the ages I have ſpoken of; and that the copies had been greatly corrupted and mutilated, by ignorant hands.

* *Bib. Latin. ab Erneſto.* Lipſ. 1774, tom. 3. p. 44.

I will

I will give one inſtance from APULEIUS, of that credulity and ſuperſtition, which, ſanctioned by antiquity, yet prevailed in the adminiſtration of remedies; and exhibits a melancholy proof of the wretched ſtate of phyſic, which, through ſo many ages, had not broke the ſhackles of druidical magic and impoſition. As a cure for a diſeaſe, called by the French *Nouè l'Equillette*, you are directed to take ſeven ſtalks of the herb *lions-foot*, ſeparated from the roots; theſe are to be boiled in water in the wane of the moon. The patient is to be waſhed with this water, on the approach of night, ſtanding before the threſhold, on the outſide of his own houſe, and the perſon who performs this office for the ſick, is alſo not to fail to waſh himſelf. This done, the ſick perſon is to be fumigated with the ſmoke of the herb *Ariſtolochia*, and both perſons are then to enter into the houſe together, taking ſtrict care not to look behind them while returning; after which, adds the author, the ſick will immediately become well.

A book under the name of MACER's *Herbal*,

Herbal, ſeems alſo to have been in common uſe in *England*, before the æra of printing. Authors do not allow it to be the production of *Æmilius* MACER quoted by OVID, but of much later date, and by ſome it is aſcribed to ODO, or ODOBONUS, a phyſician of the later times, and probably a Frenchman. This barbarous poem is in leonine verſe, and is entitled *De Naturis, Qualitatibus, & Virtutibus Herbarum*. Divers manuſcripts of it are extant in the Engliſh libraries; as, at *Cambridge*, in the *Bodleian*, *Aſhmolean*, and *Sloanean* collections.

It was tranſlated into Engliſh, as Biſhop TANNER informs us, by *John* LELAMAR, maſter of *Hereford* ſchool, who lived about the year 1373. His manuſcript is referred to as in *Sloane*'s library. Even LINACRE did not diſdain to employ himſelf on this work. "MACER's HERBAL practyſyd by "Doctor Linacro, tranſlated out of Latin "into Engliſh, London, 12mo." AMES mentions an edition of it printed in 1542; and *Palmer*, one without date, printed by *Wyre*. This jejune performance, which is written wholly on Galenic principles, treats

treats on the virtues of not more than eighty-eight ſimples.

I ſhall not detain the reader by dwelling on other authors of this claſs, whoſe names I have before recited; it will be ſufficient to obſerve, that, fettered as were the theories of this time with aſtrology, and a ſtrange mixture of the Galenic doctrine of the four elements, it extended its influence, not to the human body alone, but to all the inſtruments of phyſic. Not even a plant of medicinal uſe, but was placed under the dominion of ſome planet, and muſt neither be gathered, nor applied, but with obſervances that ſavoured of the moſt abſurd ſuperſtition.

CHAP. 3.

Manuſcripts of the Patres Botanici *ſcarce in* England—*Reſtoration of ancient knowledge, by the publications of* Pliny, Dioſcorides, *and* Theophraſtus—*The æra of commentators—Riſe of true inveſtigation by* Brunsfelſius, Tragus, Cordus, *and* Geſner—*Famous MS. of* Dioſcorides, *with illuminated figures.*

MIDDLE AGES.

AT this time manuſcripts of THEOPHRASTUS, DIOSCORIDES, and PLINY, were not only exceedingly rare throughout Europe, but thoſe of the two former were unnoticed through ignorance of the Greek language otherwiſe we cannot ſuppoſe our anceſtors could have neglected them, for the crude and barbarous works which have been mentioned. It was not till the opening of the fifteenth century, that opportunity was given to recur to theſe repoſitories of antient lore. The flight of the Greeks into *Italy*, at the ſubverſion of the Eaſtern Empire, and the ſubſequent invention

invention of printing, by bringing to light, and disseminating the purer remains of *Greece* and *Rome,* at length broke the chains of barbarism and superstition, which, during so many ages, had tyrannized over the understandings of mankind.

On this happy revolution, Botany, with other sciences, revived, and presently resumed another appearance. The publication of the *Patres Botanici* raised, at once, a spirit of emulation to investigate the subjects of their works.

PLINY was first printed, if not at *Verona,* in 1468, as is affirmed by some, and doubted by others, at least in the succeeding year, at *Venice*; and the avidity with which it was received, is manifested by the numerous impressions of it, before the end of that century.

DIOSCORIDES came forth first at *Cologn,* in a Latin translation, in 1478, and in the original, by *Aldus,* in 1495. It was afterwards published in Latin by HERMOLAUS BARBARUS and RUELLIUS, in the year 1516; by VERGILIUS, in 1518; and by CORNARUS in 1529. The learned now prefer the edition with a translation by SARACENUS, printed at *Lyons* in 1598.

THEOPHRASTUS was first printed in Greek at *Venice*, without date, and by *Aldus*, in 1495 and 1498. He was translated into Latin by GAZA in 1483, and this version has been preferred by succeeding writers.

The restoration of these sages of antiquity, immediately raised up a numerous set of commentators. Every plant was sought for, and every plant was discovered, in the works of antiquity. No drug used in medicine was esteemed *true*, unless found in DIOSCORIDES. *Scaliger* wrote animadversions on THEOPHRASTUS in 1566; in which he has corrected the version of *Gaza* in many places. Robert *Constantine* produced the parallel places in PLINY; and BODÆUS *à* STAPEL, in 1644, astonished the world, by a display of erudition on this author, in which he exhausted all farther disquisition, by the profusion of his remarks, and collations, from all preceding writers.

The commentaries on DIOSCORIDES have been more numerous. The *Corollaria* of of HERMOLAUS BARBARUS was publiſhed in 1492. To *Hermolaus* ſucceeded BRUNSFELSIUS, *Petrus Leydenſis*, LACUNA, AMATUS LUSITANUS, *Robert* CONSTANTINE, *Val.* CORDUS, and ſeveral others; and finally MATTHIOLUS, whoſe work has ſuperſeded the reſt. It was firſt printed in 1554, and paſſed through ſeventeen editions. If we may believe one of the correſpondents of this author, thirty-two thouſand copies had been ſold before the year 1561 *. The beſt edition, with the acceſſions of CASPAR BAUHINE in 1598, ſtill finds a reputable place in modern libraries.

Among the illuſtrators of PLINY, *Hermolaus Barbarus* in 1492 ſtood foremoſt. His *Caſtigationes Plinianæ*, were publiſhed in 1492, in which he ſucceſsfully corrected the text; and LEONICENUS, in the ſame year, was the firſt who employed critical knowledge on this author. The corruptions of the text afforded great ſcope afterwards to GALENIUS, RHENANUS, PIN-

* MATTHIOL. *Oper. Omn.* Ed. 1674, *in Epiſt.* p. 150.

TIANI, and others. The *Exercitationes Plinianæ* of SALMASIUS, are well known. Thoſe of the laborious and paradoxical HARDUIN, are the principal reſort of modern times.

It is a mortifying reflexion in the annals of human knowledge, that the bulk of theſe learned men, after their immenſe labours, miſtook, in numberleſs inſtances, the road to truth, and did but perplex the ſcience they wiſhed to enlighten. The deſcriptions of plants in the antient authors, were, at beſt, ſhort, vague, and inſufficient; and with this inconvenience, the ſtudy of nature herſelf was neglected. In the mean time, there aroſe a genuine ſet of cultivators, who, diſcovering this error of the commentators, ſtudied plants in the fields, where alone the beſt comments could be made. As the foremoſt of theſe, ſtands BRUNSFELSIUS. He was followed by TRAGUS, FUCHSIUS, *Val.* CORDUS, GESNER, CÆSALPINUS, and above all CLUSIUS, to whom muſt be added our own countryman TURNER. Still, even among theſe genuine reſtorers of natural knowledge, many did not act ſufficiently recollect, that all the plants of DIOSCORIDES, were not thoſe of *Europe*, but principally thoſe of *Aſia*; whilſt, inſtead of traverſing the fields of *Greece*, *Cilicia*, and the Eaſt, they were ſtraining all the deſcriptions of this author, to accommodate them to the vegetables of *Europe*. It is not ſtrange that their endeavours were but little ſucceſsful. Even, after the labours of RAUWOLF, who traverſed *Syria*, *Meſopotomia*, *Paleſtine*, and *Ægypt*, in the ſixteenth century, and thoſe of the enlightened TOURNEFORT in the preſent, it does not appear, that of the ſeven hundred plants in the *Materia Medica* of DIOSCORIDES, more than four hundred, at the fartheſt, are properly aſcertained at this time.

We learn from PLINY (lib. 25. c. 2.) that there were paintings of plants in his day; but he complains, that, through the inaccuracy of copiers, they were not to be depended on. SALMASIUS tells us, he inſpected a Greek MS. of DIOSCORIDES more than a thouſand years old, in which the plants were figured with ſufficient elegance indeed, but with little regard to truth and exact reſemblance. There are now exiſting ſeveral manuſcripts of DIOSCORIDES, with illuminated figures, particularly the famous one in the imperial library at *Vienna*, of which LAMBECIUS treats largely.

It was procured by *Buſbequius*, the emperor's reſident at *Conſtantinople*, about 1560; and is ſaid to have been copied at the expence of JULIANA ANICIA, daughter to the emperor *Flavius Anicius Olyber*, about the year 492. It has been regretted by ſome of the learned, that this MS. had not been brought earlier into *Europe*; by which means the commentators might have been ſaved much trouble. Antient, however, and ſplendid as this is, it may juſtly be doubted, whether the publication of it would have much conduced to the reſtoration of ancient Botany, and *Materia Medica*; ſince, if we are allowed to judge of the figures, from the ſpecimens copied by DODONÆUS, nothing can exceed the rudeneſs of them, or more ſtrongly juſtify the remark of *Salmaſius*. And as ſeveral of theſe are copied into GERARD's Herbal, for the ſatisfaction

ſatisfaction of the curious, I refer in the note* to ſome of theſe figures in both authors.

In juſtice, however, to theſe valuable remains, it muſt be obſerved, that, from later information, we find, there is, beſides this *Conſtantinopolitan* MS. which is in folio, another, ſuppoſed to be more ancient, in 4to. which is diſtinguiſhed by the name of *Neapolitan:* that the figures in both theſe agree extremely well; and, as *Haller* informs us, are ſufficiently exact to enable the botanical traveller, with ſuch drawings in his hands, to diſtinguiſh the plants of DIOSCORIDES in the native places of growth. It is particularly ſpecified, that the *periclymenum* of theſe manuſcripts evidently ap-

* *Coronopus.* Dod. ed. 1583. p. 109. Ger. em. 1190.
Arction. Dod. 849. Park. 1374.
Hyſſopus. Dod. 286.
Hippophaës. Dod. 373.
Aconitum Lycoctonum. Dod. 437. Ger. em. 972.
Stæbe. Dod. 123. Ger. em. 731.
Lotus Sylveſtris. Dod. 562.
Lotus Ægyptia. Dod. 563.
Tithymalus Dendroides. Dod. 368. Ger. em. 501.

pears

pears to be the *convolvulus major* of the moderns: and the *telephium*, the *cerinthe minor*. Finally, that if thoſe enumerated in the note ſo ill expreſs the plants deſigned, it muſt be wholly attributed to the fault of the copier or engraver. This intelligence is attended with regret, when we further learn, that after ſome of theſe icons were lately engraved, with a view to the publication of the whole, the deſign has been laid aſide.

I ſhall be thought, perhaps, in the foregoing pages, to have digreſſed too much. I have to allege, that a brief view of the general ſtate and progreſs of phyſic, with which my ſubject is inſeparably connected, during the dominion of the Saracens in the Eaſt, and in the ages of ignorance preceding the fourteenth century in the Weſt, ſeemed neceſſary in order to throw light on the introduction of it into this iſland. And as *England* ſhared the improvement ariſing from the reſtoration of antient knowledge, a ſhort notice of the three principal botanic authors was deemed not leſs proper.

At

At this diſtance of time, perhaps it may require ſome warmth of imagination, to picture to the mind that ſatisfaction, which ingenuous and learned men muſt have experienced, who lived when the veil was removed, which for ages had obſcured and confined thoſe elegant ſources of intellectual enjoyments, which the writings of the antients diſplay; when the means of attaining them were, by the invention of printing, ſo happily amplified, and the progreſs, not only of thoſe arts and ſciences which embelliſh, but of thoſe which alſo dignify human nature by their utility, was no longer retarded.

CHAP.

CHAP. 4.

Account of the earliest Botanical publications on the Continent—The Book of Nature—*The* Herbarius—*The* Hortus Sanitatis—*These works the basis of the* " Grete Herbal" *in* 1516; *the first Botanical publication in* England—*Account of that work*—Ascham—Copland, *both herbalists of the astrologic sect—First Botanical gardens.*

HORTUS SANITATIS.

IT was not till several years after the æra of printing, that any original work, strictly botanical, made its appearance, even on the continent; and still longer before *England* produced any publication of importance in that way.

Previous to the first dawning of this science in *England*, it is almost necessary to mention some of the productions abroad, as they were the basis of what was here first published, although, in fact, there was no original work before the *Herbal* of TURNER.

In the opinion of SEGUIER, the first book on plants, with figures, was printed at at *Augſburgh*, ſoon after the invention of wooden cuts, or tables, between the years 1475 and 1478, in the German tongue, with the title of "*The Book of Nature.*" It treats of animals and plants; of the latter, a hundred and ſeventy-ſix kinds are noticed, and many of them figured. The work is made up chiefly from PLINY, *Iſidore*, and *Platearius*.

This book ſeems to have been ſoon ſuperſeded by the famous Herbal of *Mentz*, in 1484, ſtiled ſimply "HERBARIUS;" which gave riſe, the next year, to the well-known work ORTUS SANITATIS, aſcribed to CUBA, a phyſician of *Augſburgh*, and afterwards of *Frankfort*; who, if not the author, was at leaſt the editor of an enlarged and improved edition. This work, under different editors, was the baſis of all the *Herbals* of *Europe*, for many years.

Its object is the *Materia Medica* from all nature; but vegetables occupy the greater part. The firſt edition was compriſed in four hundred and thirty-five chapters: in one, printed at Venice in 1511, which is in the black letter, they are extended to a thouſand

thouſand and ſixty-ſix; of which, one half treat on the vegetable kingdom. The author profeſſes to have drawn his reſources from HIPPOCRATES, GALEN, PLINY, AVICENNA, SERAPION, MESUES, DIOSCORIDES, PLATEARIUS, VINCENTIUS, the *Pandects*, PALLADIUS, CONSTANTIN, ALMANSER, and others. At the head of each chapter ſtands a cut, than which, ſcarcely any thing can be conceived more rude; and, in ſome caſes, nothing is more puerile or ridiculous. The pages, if printed with numbers, would amount to more than ſeven hundred. Many copies of this performance are remaining, although the *Herbarius* is become very ſcarce.

GRETE HERBAL.

Theſe books were undoubtedly the foundation of the firſt printed botanical work of any conſequence, or popularity in *England*; and which appeared under the title of "The GRETE HERBAL, with cuts;" printed for *Peter Treveris*, as *Ames* tells us, in 1516. Before the impreſſion of this book in *England*, ſome editions of the "Herbarius,"

rius," on the continent, had been augmented ſo far as to contain five hundred figures of plants. The "Grete Herbal" ſeems to have been well received in *England*, ſince there are ſubſequent copies, which bear the following dates; 1526, 1529, 1539: and in the Continuation of *Ames*, an edition is mentioned of the "Great Herbal," about the year 1550, "without the cuts." There is alſo an edition of this book ſo late as the year 1561, which is ten years after the date of TURNER'S "Herbal." That of 1526 bears the following title:

"*The* GRETE HERBALL *whiche geveth parſyt knowledge & underſtandyng of all manner of Herbes and there gracyous vertues which God hath ordeyned for our proſperous welfare & helth, for they hele and cure all manner of dyſeaſes & ſekeneſſes that fall or miſfortune to all manner of creatoures of God created, practyſed by many expert and wyſe maſters, as* AVICENNA *and other &c. And it geveth full parſyte underſtandyng of the book lately prented by me* (Peter Treveris) *named the noble experiens of the vertuous hand-*

handwarke of Surgery." *Imprynted at London in Southwarke by me* Peter Treveris, *dwelling in the Sign of the Wodows.* 1526. *the 27th day of July.*

This volume is of the ſmall folio form; and if printed with numbered pages, would make three hundred and fifty, excluſive of the Preface and Index. It includes the animal, vegetable, and mineral ſubſtances, uſed in medicine; and is ſaid in the Introduction to be "compyled, compoſed, and "auctoryſed by divers and many noble Doc- "tours and expert Mayſters in Medycynes, "as *Avicenna, Pandecta, Conſtantinus, Wil- "helmus, Platearius, Rabbi Moyſes, Johan- "nes Meſue, Haly, Albertus, Bartholomeus,* "and more other, &c."

There is no author's name to it; but there are indubitable traces of its being *fabricated* from the *Hortus Sanitatis*, and probably from the French tranſlation of that work, printed by *Caron*, at *Paris*, in 1499, with ſome alterations and additions.

It abounds with the barbarous and miſſpelt names of the middle ages, and is undoubtedly the work which TURNER refers

to

to in the Preface to his "Herbal," where he obſerves, that, "as yet there was no "Engliſh Herbal but one, al full of un-"learned cacographees, and falſely naming "of herbs."

The general order is that of the alphabet, according to the Latin names, each ſubject forming a chapter, in the whole five hundred and five; of which, more than four hundred reſpect the vegetable productions; and of theſe one hundred and fifty bear the names of plants which are natives of *England:* but the writer remarks no other diſtinction, by which they are known from the exotics. The names are given in Latin and Engliſh, but throughout the whole ſcarcely any deſcriptions. The qualities, whether *hot* or *cold, dry* or *moiſt*, according to the Galenic mode of the time, is invariably noticed, followed generally by a prolix account of the diſeaſes to which the plant is applicable, and the method of uſing it.

To each is prefixed a coarſe wooden-cut figure, as in the *Hortus Sanitatis,* from

 which,

which, on a ſomewhat ſmaller ſcale, they are evidently copied; conſiſting generally of outlines only. Each block is two inches high, and nearly as wide. Many of theſe figures are fictitious, and many miſplaced. In a variety of inſtances the ſame figure is prefixed to different plants, and in very few are they ſufficiently expreſſive of the habit, to diſcriminate even a well-known ſubject, if the name applied did not ſuggeſt the idea of it. In ſome, theſe icons are whimſically abſurd, eſpecially in the animals and minerals, being alſo copies of thoſe in the *Hortus Sanitatis*. Thoſe of the *Mandrake*, for example, exhibit two perfectly human figures, with the plant growing from the head of each; though, to do the writer juſtice, he acknowledges, that no ſuch thing exiſts in nature. At the end is ſubjoined, "an explanation of ſome terms;" and "a tract on urines."

ASCHAM.

Anthony ASCHAM, a prieſt, and vicar of *Burniſhton* in *Yorkſhire*, to which he was preferred

preferred by *Edward* VI. after a liberal education, which it might have been expected would have secured him from such delusion, gave himself up to the study of astrology, on which subject he published several tracts. He wrote also "on the Leap Year;" and the following:

"A LYTTEL HERBAL of the properties of Herbs, newly amended and corrected, with certain additions at the end of the boke, declaryng what herbs hath influence of certain starres and constellations, whereby may be chosen the best and most lucky times and days of their ministration, according to the Moon being in the signs of heaven, the which is daily appointed in the Almanack; made and gathered in the year M.D.L. xii Feb. by ANTHONYE ASCHAM, Physician." Lond. 1550. 12°.

COPLAND.

I am not able to ascertain the exact date of the underwritten, published by *William* COPLAND, a London printer.

"A Boke of the Properties of Herbs, called an Herball; whereunto is added the

"tyme that Herbes, Flowrs, and Seeds
"ſhould be gathered, to be kept the whole
"yere, with the Virtue of Herbes when
"they are ſtilled. Alſo a general Rule of
"all manner of Herbes, drawn out of the
"auncient Book of Phyſick by W. C."
London, by Wm Copland. 12mo.

BOTANICAL GARDENS.

The revival of Botany, and the conſequent eſtabliſhment of profeſſorſhips, gave riſe to Botanical gardens; a new ſpecies of luxury in horticulture, of ſingular emolument to ſcience. The hiſtory of antient gardens, hitherto not ſufficiently illuſtrated, merits the inveſtigation of the moſt learned and able writer: of the pen of a RAPIN, a MEURSIUS, a SEGUIER, or a GRONOVIUS. We learn, however, that even *Botanical* gardens are of antient date. If it may be credited, what is related of ATTALUS, the laſt king of *Pergamus*, who from his love of phyſic has been ſtiled the phyſician, he collected in his garden *hellebore, henbane, aconite,* and other poiſonous herbs, to make experiments on criminals with counter-poiſons. *Crete,* from the th earliest times renowned for the production of medicinal herbs, was the physic-garden of *Rome*. The Emperors, we are told, maintained in that island, herbarists, and gardeners, to provide the physicians of *Rome* with simples. *Castor*, a Greek, praised both by PLINY and GALEN, is said, not only to have written many volumes concerning plants, but to have had a garden at *Rome*, in which, PLINY relates, that CASTOR, at upwards of an hundred years of age, demonstrated plants, and taught him to distinguish several rare and useful species.

The utility of these institutions are self-evident. By public gardens, medicinal plants are at the command of the teacher in every lesson. By private ones, the eye, and the taste of the opulent and scientific owner, is perpetually gratified with the succession of curious, scarce, and exotic luxuries; in comparing the doubtful species, and examining them through all the stages of growth, with those to which they are allied. Add to which, that all these advantages are accumulated in a thousand objects at the same time. The first public institution of this kind, in more modern times, was that of *Padua* by the Venetians, in the year 1533. LUCAS GHINUS, the first public professor of Botany in *Europe*, was a strenuous promoter of the same designs; and by his influence procured the establishment of a garden at *Bologna*, in 1547, where TURNER himself imbibed much of that knowledge, which afterwards gave him such pre-eminence in his own country.

Among the earliest private gardens of the same kind, was that of EURICIUS CORDUS, the disciple of the venerable LEONICENUS, and of MANARDUS, two of the first commentators who displayed true Botanical criticism, on the works of the antients. CORDUS shewed himself afterwards worthy of such masters. In his *Botanologicon*, printed in 1534, he mentions his own garden, and that of NORDECIUS at *Cassel*. About the same time there were several opulent patrons of this science in *Italy*, *Germany*, and *France*, who followed this example. GESNER constructed a garden at *Zurich* in

1560;

1560; the firſt of the kind in *Switzerland.* He not only delineated plants himſelf, but maintained, at his own expence, a draughtſman and engraver, for the ſame purpoſes. TURNER appears to have had a garden for rare plants, even during his reſidence at *Cologn.* In *England* he records the garden of the duke of *Somerſet*, at *Sion Houſe*, of which he ſeems to have had the direction; and, at a later period, as hath been before obſerved, mentions alſo his own at *Wells.*

CHAP. 5.

Turner — *Anecdotes of his life* — *Account of his writings preceding the* Herbal—*His* Herbal: *the first original book of Botany, published in* England—*An account of that work*—*Contemporary Botanists mentioned by* TURNER, *as* Falconer, Wooton, Merdy, Clement—Turner's *book on baths*—Turner *not sufficiently appreciated by succeeding Botanists.*

TURNER.

THE history of English Botany to this period, from its imperfect, and even barbarous state, may perhaps not unaptly be considered as the fabulous age of the science among us. But we are now arrived at the true Era of its birth in *England*. I cannot call it the restoration, since this nation, like *Italy* in the flourishing state of *Rome*, had never been enlightened by the writings of *Greece*. It was much later before the works of those sages reached this kingdom. Manuscript copies of the PATRES BOTANICI, as hath been before observed, were

were exceedingly rare; and the language itſelf in which they are written, had made ſmall progreſs in *England*.

On this head, indeed, my ſources of information are very narrow; as far as they reach, I am not able to find, that one manuſcript of THEOPHRASTUS exiſted at this period, in any of the public libraries of *England*. Of DIOSCORIDES, there are two MSS. in the *Bodleian*, N° 3637, which bear the title of "*De Herbarum Natura et Virtutibus, cum Iconibus elegantibus*." And in the ſame collection, N° 840, an Arabic verſion of the five books, *cum Nominibus à Thoma Hyde adjectis*. Of PLINY, there is ſaid to be an entire copy in *Baliol* library, N° 279; an imperfect one, of eighteen books only, in the *Norfolk* collection, N° 2996; and an epitome, in Trinity-college, *Cambridge*, N° 459.

Even of the works of HIPPOCRATES, ſcarcely any were known except his *Aphoriſms* and *Prognoſtics*; and *Linacre* firſt made the Engliſh phyſicians acquainted with GALEN. But to return; the true Era of Botany in *England*, muſt commence

with

with Dr. *William* TURNER, who was unquestionably the earliest writer among us, that discovered learning and critical judgment in the knowledge of plants; and whose "Book of Herbs," as Dr. BULLEYN observes, "will always grow green, "and never wither as long as *Dioscorides* is "held in mind by us mortal wights." But, before I turn my attention to TURNER, I will remark, that, in an interval of thirty-four years between the first edition of "The Grete Herbal," in 1516, and that of TURNER, in 1550, I have it not in my power to refer to any publication on my subject, in the English tongue. That there were translations of several of the writers of the middle ages, has been noticed. Among those, on the continent, there were several by whose means Botany made a rapid progress. The principal were BRUNSFELSIUS, EURICIUS CORDUS, RUELLIUS, *Valerius* CORDUS, FUCHSIUS, and above all GESNER, who, possessing a genius and industry, almost unparalleled in these studies, comprehended this rising branch of knowledge, with a more expanded view than any of his predecessors,

predeceſſors, and extended its bounds beyond the limits, which, till that time, *Materia Medica* alone, had preſcribed to it. But GESNER's talents, though in Botany they were original, were ſtill more conſpicuous in his knowledge of the animal kingdom, in which, his writings will long be valued and eſteemed, by thoſe eſpecially, who, without painful reſearches, would ſee antient literature in a concentrated view. I ſpeak not of his abilities as a philologiſt and critic, in which characters he held a diſtinguiſhed place. But to proceed,

WILLIAM TURNER was born at *Morpeth* in *Northumberland*, and educated at Pembroke college, *Cambridge*, under the patronage and aſſiſtance of *Sir Thomas Wentworth*. I find him a ſtudent of that college about the year 1538, where he acquired great reputation for his learning. He applied himſelf to philoſophy and phyſic, and early diſcovered an inclination to the ſtudy of plants, and a wiſh to be well acquainted with the *Materia Medica* of the antients.

He complains of the little aſſiſtance he could

could receive in theſe purſuits. " Being
" yet a ſtudent of Pembroke hall, whereas
" I could learn never one Greke, neither
" Latin, nor Engliſh name, even amongſt
" the phyſicians, of any herbe or tree: ſuch
" was the ignorance at that time; and as
" yet there was no Engliſh Herbal, but one
" all full of unlearned cacographies and
" falſely naming of herbes."

At *Cambridge*, TURNER imbibed the principles of the reformers, and afterwards, agreeably to the practice of many others, united, to the character of the phyſician, that of the divine. He became a preacher, travelling into many parts of *England*, and propagated, with ſo much zeal, the cauſe of the reformation, that he excited perſecution from Biſhop *Gardiner*. He was thrown into priſon, and detained a conſiderable time. On his enlargement, he ſubmitted to voluntary exile, during the remainder of the reign of *Henry* VIII.

This baniſhment proved favourable to his advancement in medical and botanical ſtudies; he reſided at *Baſil*, at *Straſburgh*, at *Bonn*; but principally at *Cologn*, with

many

many other Engliſh refugees. He dwelt for ſome time at *Wieſſenburgh*; he travelled into *Italy*, and took the degree of Doctor of Phyſic at *Ferrara*. As, at this period, the learned were applying with great aſſiduity to the illuſtration of the antients, it was a fortunate circumſtance to Dr. TURNER, that he had an opportunity of attending the lectures of *Lucas* GHINUS, at *Bologna*, of whom he ſpeaks in his "*Herbal*" with great ſatisfaction; and frequently cites his authority againſt other commentators. GHINUS was the firſt who erected a ſeparate profeſſorial chair for Botanical ſcience; from whence he gave lectures on DIOSCORIDES, which he continued for twenty-eight years with great applauſe. He procured the phyſic-garden to be founded at *Bologna*, to demonſtrate the plants he ſpoke of. He was the preceptor of CÆSALPINUS and ANGUILLARA, who became two of the ſoundeſt critics in the knowledge of plants, that the age produced. TURNER reſided a conſiderable time at *Baſil*, from which place he dates the dedication of his book "On the Baths of *England* and "*Germany*."

"*Germany.*" During his reſidence in *Switzerland,* he contracted a friendſhip with GESNER, and afterwards kept up a correſpondence with him.

GESNER had a high opinion of TURNER, as appears by the following paſſage in his book *De Herbis Lunariis,* printed in 1555. "*Ante annos 15, aut circiter cum Anglicus ex Italia rediens, me ſalutaret* (TURNERUS) *is fuerit vir excellentis tum in re medica tum aliis pleriſque diſciplinis doctrinæ, aut alius quiſpiam vix ſatis memini, &c.*"

At the acceſſion of *Edward* VI. he returned to *England,* was incorporated Doctor of Phyſic at *Oxford,* appointed Phyſician to *Edward* Duke of *Somerſet,* and, as a divine, was rewarded with a Prebend of *York,* a Canonry of *Windſor,* and the Deanery of *Wells.* He ſpeaks of himſelf in the third part of his Herbal, when treating on the *herba Britannica,* which he took to be the Biſtort, as having been phyſician to the "Erle of *Embden,* Lord of *Eaſt Frieſland.*" In 1551 he publiſhed the firſt part of his hiſtory of plants, which he dedicated to the duke, his patron. His zeal in the cauſe of the reformation, which he had amply teſtified by ſeveral religious tracts, induced him to retreat to the continent, during the whole reign of *Mary*. At her deceaſe, Queen ELIZABETH reinſtated him in all his church preferments. In the dedication of the compleat edition of his "Herbal" to the queen, in 1568, after complimenting her majeſty on account of her ſkill in the Latin language, and the fluency with which ſhe converſed in it, he acknowledges with gratitude, her favours in reſtoring him to his benefices, and in other ways protecting him from troubles; having, at four ſeveral times, granted him the great ſeal for theſe purpoſes. He ſeems to have divided his time between his deanery, where he had a Botanical garden, of which frequent mention is made in his "Herbal," and his houſe in Crutched Friers, *London*. He alſo ſpeaks of his garden at *Kew*. From the repeated notices he takes of the plants in *Purbeck*, and about *Portland*, I ſhould ſuppoſe he muſt have had ſome intimate connections in *Dorſetſhire*.

Dr.

Dr. TURNER died July 7, 1568, a few months after the publication of the laſt part of his "Herbal." He left ſeveral children: his ſon *Peter* was educated to phyſic, travelled, and took degrees abroad; was incorporated doctor at *Cambridge*, and at *Oxford*; and died aged 72, in 1614; but I do not find that he inherited his father's turn to Botany.

TURNER's firſt work on the ſubject of plants, if BUMALDUS is not miſtaken, was printed at *Cologn*, under the title of "*Hiſtoria de Naturis Herbarum Scholiis et Notis vallata.*" *Colon.* apud Gymnicum 1544. 8°. *Bumaldus* is the only writer, in whom I find any mention of this book; and I ſuſpect, it was not republiſhed in *England*. It was followed by a ſmall volume under the title of "NAMES OF HERBES, in Greek, Latin, Engliſh, Dutch, and French." *Lond.* 12°. 1548. This nomenclator is, I believe, become very ſcarce; ſince it has not yet found its way into the copious and magnificent collection of Sir *Joſeph* BANKS.

Dr. TURNER's knowledge in natural hiſtory was not confined to Botany; his earlieſt publication

publication appears to have been, a treatiſe on birds, under the following title:

"*Avium præcipuarum quarum apud Plinium et Ariſtotelem mentio eſt brevis et ſuccincta hiſtoria, ex optimis quibuſque ſcriptoribus contexta. Scholio illuſtrata et aucta. Adjectis nominibus Græcis, Germanicis, et Britannicis.*" Coloniæ 1543. 8°. Not having ſeen this volume, I can only ſay, that TURNER is mentioned by his friend GESNER, in reſpectful terms, as an ornithologiſt. "*Avium quidem nomina et naturas ante nos et pauci et breviter attigerunt ex quibus Gyb. Longolius Germanus, et Gulielmus* TURNER *Anglus viri doctiſſimi præcipuam merentur laudem.*" *Geſn. Præf. ad Avium Hiſt.* TURNER alſo contributed to enrich GESNER's muſeum (the firſt collection of that kind,) with natural curioſities, which he ſent from *England.* To which I add, that Dr. MERRET gives the following teſtimony to the worth of TURNER, in the Preface to his "Pinax:" "*Conſului in quibuſdam* TURNERUM *noſtratem inter viros ſuæ ætatis exercitatiſſimum qui librum de avibus edidit mole. parvum at judicio majorem.*"

Prefixed to the third volume of the *Frankfort* edition of GESNER's *Historia Animalium*, in 1620, we find a letter from Dr. TURNER, relating to the English fishes; which sufficiently proves, that he had no inconsiderable degree of knowledge in that part of zoology. He makes an apology for the imperfections of it, as being written from memory, and at a distance from all his notes and observations. It consists of three pages, in which he has briefly described more than fifty species; and it seems to be intended principally to give GESNER information on the English names, which TURNER has carefully noted, and often added the provincial appellations. He takes in both sea and river fish, and includes also the scallop and the cockle. This letter was written from *Weissenburgh*, and is dated Nov. 1, 1557. He undoubtedly pursued this branch of zoology much farther; since it appears from his dedication to the queen, that he intended "to set out a book of the names "and natures of the fishes of her majesty's "realms."

But the work which secured his reputation to posterity, and entitled him to the

character

character of an original writer on that subject, in *England*, is his "Hiftory of Plants," printed at different times, in three parts, in folio, with cuts. The firft at *London*, in 1551, under this title, "A NEW HERBALL, wherein are contayned the names "of herbes in Greeke, Latin, Englifh, Duch, "Frenche, and in the Potecaries and Herbaries Latin, with the properties, degrees, "and natural places of the fame gathered. "For Steven Mierdman." *Lond.* 1551. The fecond part at *Cologn*, 1562, during his exile in the reign of *Mary*. With this was reprinted the firft part; and his "Book "on the Bathes of *England* and *Germany*."

In 1568 thefe were reprinted, with the addition of the third part, which bears the following title: "The third part of Wm TURNER'S HERBAL, wherein are contained the herbes, rootes, and fruytes, whereof is no mention made of *Diofcorides*, *Galene*, *Plinye*, and other old authors. Imprinted at *Collen*, by Arnold Birckman, in the year of our Lord 1566." The dedication, however, to the company of furgeons, is dated from *Wells*, June 24, 1564.

Dr. TURNER's "Herbal" is printed in the black letter, agreeably to the general uſage of the times, and is embelliſhed with the figures of moſt of the plants he deſcribes.

The arrangement is alphabetical, according to the Latin names; and, after the deſcription, he frequently ſpecifies the places of growth. He is ample in his diſcrimination of the ſpecies, as his great object was, to aſcertain the *Materia Medica* of the ancients, and of DIOSCORIDES in particular, throughout the vegetable kingdom. To this end he beſtows much criticiſm on the commentaries of FUCHSIUS, TRAGUS, MATTHIOLUS, and other of his contemporaries; and profeſſes to have corrected many of their miſtakes, in the application of the names of DIOSCORIDES. In all this he has ſhewn much judgment, and, I may add, much moderation, in avoiding, more than uſual, the licence taken by many of the commentators, of applying the names of plants deſcribed in THEOPHRASTUS, DIOSCORIDES, and PLINY, to thoſe of the weſtern parts of *Europe*. What he ſays of the virtues of plants,

plants, he has drawn from the ancients; but has, in numberleſs inſtances, given his opinion of their qualities, in oppoſition to thoſe ſages, and recorded his own experience of the virtues. He no where takes any doubtful plants upon truſt, but appears to have examined them with all the preciſion uſually exerciſed at a time when method, and principles now eſtabliſhed, were unthought of; every where comparing them with the deſcriptions of the antients and moderns. He firſt gave names to many Engliſh plants; and, allowing for the time when ſpecifical diſtinctions were not eſtabliſhed, when almoſt all the ſmall plants were diſregarded, and the *Cryptogamia* almoſt wholly overlooked, the number he was acquainted with, is much beyond what could eaſily have been imagined, in an original writer on his ſubject.

The third part of his "Herbal," dated from *Welles*, June 24, 1564, he dedicates to the company of ſurgeons; and apologizes for its imperfections: "Being ſo much "vexed with ſickneſs, and occupied with "preaching, and the ſtudy of divinity, and

 "exerciſe

" exercise of discipline, I have had but
" small leisure to write Herballes."

In this part, he professes to treat on the plants not known to DIOSCORIDES and the antients. It consists of near an hundred articles, among which we find introduced many of the exotic subjects, which had before been but little known; such as *cassia fistula*, *cubebs*, *guaiacum*, *nutmegs*, *myrobalans*, *nux indica*, *nux vomica*, *anacardium*, *rhubarb*, *sarsaparilla*, *senna*, and *tamarinds*. For these, many new figures were cut, which are executed in a stile superior to the others. The remainder are principally the productions of our own country.

The compleat edition of TURNER'S "Herbal," in 1568, was printed at *Cologn*, unquestionably to receive the advantage of the figures, probably at that time the property of *Birkman* the printer. They are the same with which the octavo edition of FUCHSIUS was first printed in 1545; in all five hundred and twelve. Of these, TURNER has used upwards of four hundred; to which he has added about ninety new, making the whole number five hundred and two.

There

There are ſome inſtances of the wrong application of theſe figures; an error that might readily happen, when the author was at ſuch a diſtance, and was common in almoſt all ſimilar works of that time. There are alſo ſeveral figures to which no deſcription of the plants can be found; for inſtance, the ſix figures of the *Geraniums* from FUCHSIUS occur, with a ſlight mention of only two ſpecies in the text.

TURNER is the firſt author who has given a figure of the Lucern; which, I apprehend, he firſt brought into *England*, and named *Horned Clover*. He treats largely of its cultivation, from PLINY, PALLADIUS, and COLUMELLA.

In the dedication to the firſt edition of his "Herbal," in 1551, Dr. TURNER ſpeaks in very reſpectful terms of the botanical knowledge of ſeveral of his contemporaries; and apologizes for his undertaking ſo arduous a matter, while there were learned Engliſhmen better qualified. He enumerates Dr. CLEMENT, Dr. MERDY, *Owen* WOOTON, and Maſter FALCONER.

The laſt-mentioned is ſeveral times introduced in the body of the work. I can ſcarcely doubt that he was *John* FALCONER, who is recorded as having communicated many Engliſh plants to AMATUS LUSITANUS, who taught phyſic at *Ferrara* and *Ancona*, and made himſelf known as a commentator on DIOSCORIDES in 1553. In treating on the *Glaux*, of which TURNER gives a new figure, he ſays, "He never ſaw it in *England*, except in Maſter *Falconer*'s book; "and that he brought it from *Italy*." From this and other like citations, it may reaſonably be conjectured, that "*Falconer*'s Book" was an *Hortus Siccus*; and if ſo, muſt have been among the earlieſt collections of that kind, that is noticed in *England*.

In appreciating the merit of Dr. TURNER as a Botaniſt, due regard muſt be had to the time in which he lived; the little aſſiſtance he could derive from his contemporaries, of whom, BRUNSFELSIUS, RUELLIUS, FUCHSIUS, and TRAGUS, when he publiſhed his firſt part of the "Herbal," were the chief; in which view, he will appear

pear to have exhibited uncommon diligence and great erudition, and fully to deſerve the character of an original writer.

Our author paid early attention to mineral waters. He was probably the firſt who wrote on the baths of *Bath*, in *Somerſetſhire*. He viſited ſeveral of the mineral ſprings in *Germany*, *Switzerland*, and *Italy*; and drew up, whilſt abroad, a ſhort account of ten of thoſe waters; to which he prefixed a more enlarged hiſtory of the waters of *Bath*. This was written, as it ſhould ſeem, at *Baſil*, and is dedicated to his "well-"beloved neighbours of *Bath*, *Briſtow*, *Wells*, "*Winſam*, and *Charde*," March 10, 1557. He adjudged the principle of Bath water to be brimſtone, and poſſibly a little copper, from the vicinity of that metal in the neighbouring mountains. He ſays, he had been informed, that, beſides brimſtone, the King's bath held alum, and the Croſs bath ſaltpetre; but that he could find neither. He concludes his account of the baths, by a ſet of general rules for all who drink mineral waters; many of which do him no diſcredit,

dit, when compared with the injunctions of modern physicians.

Our author also wrote "On the Nature "of Wines commonly used in *England*," in vindication of the use of Rhenish wines. To this was annexed a tract "On the Na- "ture and Vertue of Treacle." But, as I never saw these treatises, I can give no account of them.

Dr. TURNER was the author of many polemical and religious treatises, chiefly written in defence of the Reformation. Of these, a list is given in the *Athenæ Oxonienses*, and a more accurate and enlarged one in Bishop TANNER's *Bibliotheca*. Several of his tracts are yet in manuscript, in various libraries. He collated the translation of the Bible with Hebrew, Greek, and Latin copies, and corrected it in many places.

He procured to be printed at *Antwerp*, a new and corrected edition of the *Historia Gentis nostræ*, f. *Angliæ*, written by *William of Newburgh*, from a manuscript he found in the library of *Wells*; but complains, that the printer not only omitted to insert certain

tain articles ſent by him, but left out the preface he ſent him, ſubſtituting one of his own. Our author alſo tranſlated ſeveral works from the Latin, particularly "The "Compariſon of the Old Learning and "the New;" written by *Urbanus Regius*. Southwark. 1537. 8°; and again 1538 and 1548.

I will not conclude this ſhort memoir of Dr. TURNER, without remarking, that the ſucceeding Herbaliſts, GERARD, JOHNSON, and PARKINSON, ſeem not to have paid due honour to his merit and learning, from the ſilence they obſerve relating to him in their writings. GERARD, indeed, mentions in his Preface, "that ex-"cellent work of maſter Dr. TURNER;" and, in another place, ſtiles him "that ex-"cellent, painefull, and diligent phyſition, "Mr. Dr. TURNER, of late memorie." In juſtice to TURNER, they ſhould have noticed all the plants he has recorded, particularly the natives of *England*.

RAY, at the diſtance of near a century, was ſenſible of his worth, having ſtiled him

him " a man of ſolid erudition and judg-
" ment *."

* In honour of TURNER, his name has been annexed, by *Plumier*, the French Botaniſt, to a new genus of plants, well known at this time in the Engliſh gardens. It was firſt diſcovered by SLOANE, in *Jamaica*, and deſcribed by him under the title of *Ciſtus Urticæ folio*.

CHAP. 6.

Dr. Bulleyn—*Anecdotes of his life—His Herbal; or Book on Simples—His Defence of the Fertility of* England.

Dr. Thomas Penny: *Short Anecdotes of—The friend and correſpondent of* Geſner, Cluſius, *and* Camerarius.

Maplet—Morning.

BULLEYN.

CONtemporary with TURNER lived Dr. *William* BULLEYN. Although this writer does not come ſtrictly within my plan; yet, as he lived at a period barren of intereſting materials, and, as we learn from him ſeveral curious anecdotes reſpecting natural hiſtory and the ſtate of gardening in England at that period, he cannot be paſſed over in ſilence.

Biſhop TANNER briefly notices Dr. BULLEYN, and his writings; but his life is amply written in the *Biographia Britannica*, to which I muſt principally be indebted for my information.

He

He was born in the Iſle of *Ely*, in the early part of *Henry* the Eighth's reign, and was educated at *Cambridge*, though, as *Wood* ſays, he afterwards reſided ſome time at *Oxford*. It appears that he had travelled over ſeveral parts of *Germany*; that he viſited *Scotland*, and had taken many tours in his native country; in all which, he ſtudied the natural productions with a zeal and ſucceſs not common in that age. In an early period of his life, he was much converſant about the city of *Norwich*. In June 1550, he was inſtituted to the rectory of *Blaxhall*, in *Suffolk*, where his relations reſided. This preferment he reſigned in 1554. Where he took the degree of doctor in phyſic, is not aſcertained; but, from his prior attachment to phyſic, his known oppoſition to the doctrine of Tranſubſtantiation, and the reſignation of his living in the beginning of *Mary*'s reign, it may be fairly conjectured, that he did not take his degrees in that faculty till after that period, and probably abroad. After this, we find him removed to the city of *Durham*, where he practiſed phyſic, and became poſſeſſed of

property

property in the ſalt-pans, near *Tinmouth Caſtle*. On the death of his patron, Sir Thomas *Hilton*, he removed to *London*, where he became a member of the college of phyſicians, and acquired reputation as a phyſician, and a man of learning. This event took place about the year 1560. He had the misfortune to loſe great part of his library, with his manuſcript upon "Healthfull Medicines," by ſhipwreck; and after this diſaſter, met with moſt unjuſt and malevolent treatment, from a brother of Sir Thomas *Hilton*, by whom he was accuſed of having murdered his late patron, who died, in fact, of a malignant fever. And although his innocence was fully manifeſted, yet his enemy perſiſting further in his perſecution, found means to throw him into priſon, for debt, where he wrote a great part of his medical treatiſes. He died Jan. 7, 1576. He appears to have been much attached to the principles of the reformation. Biſhop TANNER ſays he was a man of acute judgment and true piety.

I am not acquainted with any print of Dr. TURNER. Of Dr. BULLEYN there is a profile with a long beard, before his "Government of Health," and a whole length of him in wood prefixed to the "Bulwarke of Defence;" which book is a collection of moſt of his works. He was an anceſtor of the late Dr. STUKELY, who, in 1722, was at the expence of having a ſmall head of him engraved.

The part of his works, which has the neareſt connection with my ſubject, is in his "Bulwark of Defence," in fol. 1562.

It is entitled, "A Book of Simples, be-"ing an HERBAL in the form of a dia-"logue, at the end of which are the cuts "of ſome plants in wood." In this piece he obſerves, that *tormentil*, in paſtures, prevents the rot in ſheep; and adds, that the fact was confirmed by the ſhepherds in ſundry parts of Norfolk. In his enumeration of the virtues of ſimples, from other authors, he does not fail to record his own experience on the power of ſeveral, in removing ſevere diſeaſes. Of the effects of Dittander, *calamus aromaticus*, the Daiſy, and others, he adduces particular inſtances. It were to be wiſhed, that ſucceeding obſerva-

tions,

tions, had confirmed his reprefentation. His travels, and the great attention he had paid to the native productions of his own country, had given him a comprehenfive view of the natural fertility of the foil, and climate of *England*; which, from the tenour of his writings, feems to have been, at that time, by fome people much depreciated. He oppofes this idea with patriotic zeal and concern, and alleges various examples, to prove, that we had excellent apples, pears, plums, cherries, and hops, of our own growth, before the importation of thefe articles into *England* by the *London* and *Kentifh* gardeners, but that the culture of them had been greatly neglected. He endeavours to confirm the natural fertility of the land, from the memorable inftance of the *fea peafe*, on the beach, near *Orford* and *Aldborough*; by an immenfe crop of which the poor were preferved in a time of dearth, in the year 1555. Of which fee further accounts in *Johnfon's* GERARD, p. 1250; PARKINSON'S "*Theatre*," p. 1060; and LOBEL'S *Illuftrationes*, p. 164.

To conclude, Dr. Bulleyn's ſpecific knowledge of Botany ſeems to have been but ſlender. His zeal for the promotion of the uſeful arts of gardening, the general culture of the land, and the commercial intereſts of the kingdom, deſerved the higheſt praiſe, and for the information he has left of theſe affairs, in his own time, poſterity owe him acknowledgments.

Although the progreſs of gardening does not enter into my plan, yet I am tempted, in this place, to remark, that, notwithſtanding culinary herbs and roots, and many fruits, are ſaid to have been imported in the reign of HENRY *the Eighth*, from *Holland* and *France*; and that the true æra of improvement in this art, cannot be carried, at the moſt remote time, beyond the ſame reign, yet it may juſtly be doubted, whether it was then in ſo low a ſtate as hath been uſually repreſented. With other arts, in its progreſſion weſtwards, that of Horticulture muſt be ſuppoſed to have reached the *Low Countries* and *France*, before *England*; and a general, and prior ſuperiority to our neighbours may be granted; and that a faſhion,

faſhion, and a too great fondneſs for rarities of foreign growth, might influence the London market, of which the ſpirit of commerce would not fail to take advantage, muſt likewiſe be admitted. But, to the arguments and proofs alledged by *Dr.* BULLEYN, in defence of the fertility of his native ſoil, and the perfection of our own products; and, as a proof of the ſucceſsful cultivation of thoſe times, I add, that from an inſpection of our old Herbals, and particularly of PARKINSON's *Paradiſus*, we find the various ſpecies of culinary herbs, roots, and of fruits, multiplied in *England* to ſuch a variety, as implies a preceding courſe of culture carried on for a ſeries of time, inconſiſtent with that poverty of produce which hath been ſurmiſed.

PENNY.

Having introduced to the reader, the two firſt reſpectable writers on Botany in *England*, I cannot but regret my want of ſufficient information, to reſcue from an almoſt total obſcurity, the name of Dr.

Thomas PENNY, an Engliſhman of the ſame age; who, although not an author himſelf, was indubitably a man of great attainments in the natural hiſtory, and eſpecially in the Botany, of his time. GERARD ſtiles him "A ſecond *Dioſcorides*, for his ſingular "knowledge in plants." I cannot aſcertain the date of his birth. It appears that he was a fellow of the royal college of phyſicians, and that he had travelled into various parts of *Europe*. He had reſided in *Switzerland*, and had viſited, if not made ſome ſtay in, the iſland of *Majorca*. That he had diligently ſearched both the northern and ſouthern parts of *England* is manifeſt, from the variety of rare plants diſcovered by him, and communicated to LOBEL and GERARD. He was perſonally known to GESNER and CAMERARIUS, and afterwards frequently ſupplied them with rare plants, for their reſpective *Herbaria* and gardens.

During his reſidence in *Switzerland*, he collected many plants of that country, and from the confines of *France*. He aſſiſted GESNER, as appears by his obſervations and

and animadverſions on that author's tables, publiſhed by SCHMIEDEL from the collections of TREW, in 1753, in which the moſt honourable teſtimony is given to his abilities. I ſuſpect he was in *Switzerland*, at the time of GESNER's death, and aſſiſted WOLF in arranging the plants, and memorials of their deceaſed friend.

There can be no doubt that PENNY and CLUSIUS were alſo perſonally acquainted. They appear to have had a ſtrict intimacy, and the latter was obliged to PENNY for a variety of curious articles inſerted in his *Rariores*, and in the *Exoticæ*. Dr. PENNY brought from *Majorca* the *hypericum balearicum*, which CLUSIUS named *myrtocistus* PENNÆI after him, as he did a gentian, now the *ſwertia perennis*. The ſame of the *geranium tuberoſum*. The *cornus herbacea*, that beautiful native of the *Cheviot* hills, was firſt revealed to the curious by this induſtrious naturaliſt.

Dr. PENNY's acquirements in natural hiſtory extended beyond the knowledge of plants. He is one of the firſt Engliſhmen whom I have met with, who had ſtudied insects. There are letters witten by him to CAMERARIUS, in the year 1585, preserved in TREW'S collections, which prove his knowledge in *entomology*, to have been extensive in that day: and it is supposed by SCHMIEDEL, that GESNER'S drawings of *Papilio*'s, passed into the hands of PENNY. This supposition is rendered more probable, when it is recollected, that the *Theatrum Insectorum* of MOUFET, was a work begun by *Dr. Edward* WOOTON, *Conrade* GESNER, and Dr. PENNY, and received only the finishing hand from MOUFET.

Dr. PENNY died in 1589, and is said by JUNGERMAN to have left his papers to MOUFET and TURNER; but, in this account there is surely a very striking anachronism, since TURNER himself died in the year 1568.

MAPLET.

John MAPLET, master of arts, of *Cambridge*, published in the year 1567, "A "GREEN FOREST; or, Natural His-"tory; wherein may be seen, the sove-"raign

" raign vertues of all kinds of ſtones, and " metals, *herbs*, *trees*, beaſts, fouls, and " fiſhes; 112 leaves, 8°." I have not ſeen *Maplet*'s book; but from the title of another work of his, " The Dial of Deſtinie; " or, Influence of the Seven Planets over " all Kinds of Creatures here below," publiſhed in 1581, it may fairly be preſumed, that he was deep in the fancies of the aſtrologic ſect.

MORNING.

Between the publication of TURNER's Herbal, and that of LYTE, I find a book, of which, not having ſeen it, or been able to refer to any account, I can only recite the title. " The Treaſure of Euonymus by " *Peter* MORNING; with wooden cuts. " Imprinted by John Day." 4°, 1575.

CHAP. 7.

Lyte—*Anecdotes of—Not an original writer in Botany—His Herbal a translation from* Clusius's *version of* Dodoens—*Small accession made to* English *Botany by this work.*

LYTE.

HENRY LYTE, Esq; of an ancient family, at *Lytes-Carey*, in *Somersetshire*, was the next after TURNER who published an *English Herbal*. He was born in 1529, and became a student at *Oxford* in the latter end of *Henry* VIII. about the year 1546. He afterwards travelled; and at length retired to his patrimony, where, as *Wood* says, "by the advantage of a good foundation "of literature made in the university and "abroad, he became a most excellent scho-"lar in several sorts of learning." He was the author of various publications of the historical kind, which are enumerated in the *Athenæ Oxonienses*. He died at the age of 78, and

and was buried at *Charlton-Mackerel*, in the ſame county. He left a ſon, who drew up a genealogy of *James* I. for which the king rewarded him with his picture in gold, ſet with diamonds; and the prince, afterwards *Charles* I. gave him alſo his picture in gold.

Although Mr. LYTE does not rank among original writers in Botany, his work nevertheleſs ſeems to have been well received. Even the arrangement alone would inſtantly give it a great advantage over *Turner*'s book. It is profeſſedly a tranſlation from the French verſion of the Dutch Herbal of DODOENS, written by the author in 1553, and tranſlated by *Cluſius* in 1557; being the firſt of his publications. Of DODOENS, it will be neceſſary to give ſome account; but I ſhall defer it till I ſpeak of GERARD, as the improved editions of DODOENS's book were the baſis of that author's work.

The firſt edition of LYTE's Herbal was publiſhed at *Antwerp*. It is printed in the black letter, and bears the following title:

" A NIEWE HERBALL, or HISTORIE OF

" PLANTES,

" PLANTES, wherein is contayned the whole
" diſcourſe and perfect deſcription of all ſorts
" of herbs and plantes; their divers and ſun-
" dry kindes; their ſtraunge figures, faſhions,
" and ſhapes; their names, natures, and ope-
" rations and vertues: and that not only of
" thoſe which are here growyng in this our
" countrie of Englande, but of all others alſo
" of forayne realmes, commonly uſed in phy-
" ſicke. Firſt ſet forth in the Doutche or
" Almaigne tongue, by that learned D. *Rem-*
" *bert* DODOENS, phyſition to the emperor;
" and now firſt tranſlated by

" *Henry* LYTE, *Eſquyer*.

" At London, by me, *Gerard Dewes*. 1578."
—The Colophon, " imprinted at *Antwerpe*,
" by me, *Henry Loe*, book-printer." pp.
779.

Mr. LYTE dedicates his work to queen *Elizabeth*; and has prefixed the preface and appendix in Latin, from DODOENS, or DODONÆUS. The latter of theſe is a collection from DIOSCORIDES and CATO, but chiefly from PLINY, relating to the riſe and progreſs of botanical and agricultural knowledge

knowledge among the Romans; and in commendation of gardens, with rules for laying them out, and managing them to advantage.

He has followed his original in dividing his subjects into six books; and, although the general arrangement is confused, LYTE has the merit of having introduced a particular order in each chapter, or genus, much superior to that of TURNER; having divided the species, description, place, time, names, nature, and virtues, under these several titles, into distinct sections. This arrangement was adopted by GERARD and PARKINSON.

LYTE describes one thousand and fifty species, of which eight hundred and seventy are figured. The blocks are, I believe, the same with which CLUSIUS's own translation was printed; being, as far as those extend, copies from the octavo edition of FUCHSIUS. Most of TURNER's figures are found in LYTE. The remainder are such as had been cut for the subsequent works of DODOENS, and afterwards embellished the *Pemptades* of that author, and GERARD's history.

hiſtory. The Engliſh tranſlator added about thirty new ones. Among theſe, ſeveral are in a ſtyle ſuperior to thoſe of CLUSIUS and GERARD; ſuch are particularly, the *Salvia Æthiops*; the *Stratiotes aloides*; the *Rha*, or *Centaurea Rhaponticum*; and others.

Some are original: I cite only the *Erica Tetralix*, of which I find no figure prior to *Lyte*'s; that of GERARD (or, which is the ſame, of *Cluſius)* applied to it by JOHNSON, being certainly intended to repreſent another ſpecies, and is accordingly referred to the *Mediterranea* by LINNÆUS.

The firſt edition of *Lyte* is adorned with a finely-cut impreſſion in wood of DODOENS, in the thirty-fifth year of his age; and a large engraving of Mr. *Lyte*'s coat of arms.

This firſt edition was undoubtedly printed at *Antwerp*, to receive the advantage of the figures. The ſubſequent editions, therefore, afterwards printed in *England*, are without figures. It was reprinted, as *Ames* informs us, in 1586, and in 1595; and, according to *Wood*, by *Ninion Newton*, at *London*, in 1589, in quarto alſo, without cuts.

cuts. I find editions mentioned, with the dates 1600 and 1619, which, if genuine, and not in the title-page only, is a proof of its popularity; and that it was not superseded by the larger work of GERARD in 1597. SEGUIER even quotes one, so late as the year 1678.

As in the interval between the publication of CLUSIUS's French translation in 1557, and the English version of it by LYTE in 1578, the author had at different times compleated the several parts of his *Historiæ Plantarum*, it may be presumed, that LYTE profited by those works. From some of the commendatory verses prefixed, it should seem, that *Dodoens* himself communicated additions to LYTE. As I have not had an opportunity of comparing the French version of *Clusius* with LYTE, I cannot notice the nature of his alterations, or the extent of his additions. The introduction of the English names was a necessary augmentation.

In the mean time, there seems to be no ground for the criticism of THRELKELD; who accuses LYTE of having omitted the

Purgantium

Purgantium Hiſtoria of DODONÆUS, of which LYTE appears unqueſtionably to have introduced the moſt material ſubjects.

Engliſh Botany, however, received little or no acceſſion from LYTE himſelf. It is not in more than about twenty inſtances, that he has even pointed out the local ſituation of any rare Engliſh plants; and, in theſe inſtances, there is ſcarcely one, which had not been thus *ſpecifically* recorded by TURNER and LOBEL, before him.

Hence, I am not able to give LYTE the credit, although he lived at ſo early a period, of being the firſt diſcoverer of a ſingle ſpecies of rare growth. Yet, as it is but juſtice to ſuppoſe him well acquainted with all the common plants, ſo a large number of theſe, which had been unnoticed by TURNER, or are not eaſily aſcertained in his work, will be found firſt announced to the Engliſh Botaniſt in LYTE. I confeſs, however, that it is extremely difficult to determine, in a variety of inſtances, whether the general places of growth, as mentioned in this author, are inſerted from his own knowledge, or whether they ſtand as tranſlated by him from CLUSIUS. It is this doubt that has induced me, not unfrequently, to aſcribe to GERARD, or JOHNSON, the firſt knowledge of many common plants certainly aſcertained by them, that occur, nevertheleſs, in LYTE's work.

This author furniſhes very few obſervations which tend to illuſtrate the ſtate of the ſcience, between the time of TURNER and his own. Nor does he mention, in more than one or two inſtances, any of his contemporaries. Under the article *Verbaſcum*, he ſpeaks of "the pleaſant garden of "*James Champaigne*, the deer friende and "lover of plantes:" but without any information of his character, or place of abode: And, under that of *Sweet Trefoil*, "the "garden of maiſter Rich."

CHAP.

CHAP. 8.

Lobel—*Anecdotes of—Of* Flemiſh *extraction, but lived chiefly in* England—*Travelled with* Lord Zouch—*Entitled Botaniſt to* King James—*The* Adverſaria, *written jointly by him and* Pena—Lobel *a learned man, and well verſed in the* Materia Medica — Engliſh Botany *greatly augmented by him—Promoters of Botany and gardening mentioned by him.*

Newton—*His* Herbal *to the Bible—only a tranſlation from* Lemnius.

LOBEL.

MATTHIAS de LOBEL, though not a native of *Britain,* contributed ſo largely to the emolument of Engliſh Botany, that he juſtly claims attention in the object of this work. LOBEL was of Flemiſh extraction, and was born in 1538 at *Liſle,* where his father was in the profeſſion of the law.

He informs us, that; at the age of ſixteen, he was enamoured with the love of plants;

plants; and had an unconquerable desire to know the names and properties of those used in physic. He studied at *Montpelier*, under the famous RONDELETIUS. During his residence there, he travelled over the south of *France* in search of simples.

At *Narbone* he formed a connection with *Peter* PENA, who was jointly concerned with him in his first work the *Adversaria*. On leaving *France*, he extended his researches by travelling over *Switzerland*, the county of *Tyrol*, some parts of *Germany*, and *Italy*; and on his return settled as a physician at *Antwerp*, and afterwards at *Delft*. He was then made physician to *William* Prince of *Orange*, and to the States of *Holland*. On what occasion he removed into *England*, or at what period of his life, I cannot ascertain. From the circumstance however of the *Adversaria* bearing date at *London* in 1570, it should seem to have been before that time, which opinion is somewhat corroborated, by his informing us, that Dr. TURNER had given him, "long before," the seeds of the *sea kale*.

In *England*, he obtained the patronage of

Lord Zouch, whom he attended in 1592, in his embaſſy to the court of *Denmark*. This tour furniſhed him with further means of augmenting his knowledge in Botany; and, through the correſpondence he formed there, of introducing into *England* ſeveral exotic rarities, before that time unknown to this country. He had the ſuperintendance of a garden at *Hackney*, which he calls a phyſic-garden, cultivated at the expence of his patron. He was afterwards ſtiled Botaniſt to King *James*, as appears by the *imprimatur* to the ſecond edition of the *Adverſaria*; and by his own letter prefixed to GERARD's "Herbal." Whether any emolument was annexed to this title, I am unable to decide. He had a daughter married to a Mr. *James* COEL, who lived at *Highgate*, near *London*; and it is probable, from the very frequent mention that LOBEL makes of that place in his laſt work, the *Illuſtrationes*, that he reſided in the latter years of his life with his ſon-in-law.

He died in 1616, aged 78. There was a print of LOBEL, but it is very ſcarce, I have only ſeen it in the collection of the late Mr. *Gulſton*.

The firſt of LOBEL'S publications, and which more eminently agrees with the deſign of this work, as it brought a large acceſſion to Engliſh Botany, was the *Stirpium Adverſaria*. The profeſſed intention of this work was to inveſtigate the Botany and *materia medica* of the antients, and particularly of DIOSCORIDES; and LOBEL is judged to have corrected the errors of MATTHIOLUS, upon that author, in many inſtances.

As PENA was jointly concerned with LOBEL in this work, it is become impoſſible, at this time, to aſſign to each their ſeparate ſhare. The firſt edition of the *Adverſaria*, dated at *London* 1570, was dedicated to the queen. This dedication was omitted in an edition printed at *Antwerp* in 1576. Editions bearing date 1571, 1572, are recorded, but it may be doubted whether theſe were more than title-page alterations. To that of the whole *Adverſaria*, which bears date London 1605, by *Purfoot* alſo, is prefixed LOBEL'S *Animad-*

verſiones in Rondeletii methodicam Pharmaceuticam officinam; containing 156 pages. After this, the title, and a dedication to the profeſſors at *Montpelier*, printed by *Purfoot*; but the ſucceeding firſt part of the *Adverſaria*, is on a much better paper, and in a finer type, and evidently printed by *Plantin* as far as to page 450; to which ſucceeds one leaf, added in *Purfoot*'s type, containing the account of the *Plocamos* of *Portland*, and of the *Barnacle*, the fabulous hiſtory of which he relates, without wholly denying it. Then follows, (the pages being continued,) the ſecond part of the *Adverſaria*, now firſt printed by the *London* printer. To which is annexed, LOBEL's "Tract on the Balſams, Cinnamon, Caſ-"ſia," and various other matters; with a ſmall treatiſe on the dropſy, and the *elephantiaſis*, written by his much reverenced maſter RONDELETIUS.

The ſecond edition bears the following title, "*Dilucidæ Simplicium Medicamentorum explicationes, et* STIRPIUM ADVERSARIA, *perfacilis veſtigatio, luculentaque acceſſio ad priſcorum, præſertim Dioſcoridis et recentiorum*

Materiæ

Materiæ Medicæ ſolidam cognitionem. Methodo exquiſitiſſima, a notioribus ſummiſque claſſium generibus ad ultimas uſque ſpecies digeſta. Authoribus Petro PENA, *et Matthia de* LOBEL *medicis. Quibus acceſſit* ALTERA PARS, *cum prioris illuſtrationibus, caſtigationibus, auctariis, rarioribus Plantis. Selectioribus remediis, ſuccis medicatis et metallicis, medicinæ theſauris, opii opiati antidoti, decantatiſſimique chymiſtarum et germanorum laudani opiati formulis. Opera et Studio Matthiæ de Lobel, Londini* 1605. *pp.* 549.

Acceſſit Matthiæ de Lobel, in Rondeletii Methodicam Pharmaceuticam animadverſiones cum Myrei paragraphis. pp. 156."

Reprinted at Frankfort in 1651.

In the execution of this work, there is exhibited, I believe, the firſt ſketch, rude as it is, of a natural method of arrangement; which, however, extends no farther than throwing the plants into large tribes, families, or orders, according to the external appearance, or habit of the whole plant or flower; without eſtabliſhing any definitions or characters. The whole forms forty-four tribes. Some con-

 tain

tain the plants of one, or two modern *genera:* others many; and ſome, it muſt be confeſſed, very incongruous to each other. On the whole, they are much ſuperior to DODOENS's diviſions; and ſufficiently teſtify, that the author was ſenſible of the want of a better arrangement than the mere alphabetic order, or that formed from the ſuppoſed qualities, and uſes in medicine.

At the head of each tribe, or family, he prefixes a ſynoptical view of all the ſpecies to be deſcribed under it. His method, then, is to give the *Greek* and *Latin* name; and, wherever he can, the name of the genus and ſpecies, in *German*, *Dutch*, *French*, and *Engliſh*. Then the deſcription of the plant, the time of flowering, the country in which it grows ſpontaneouſly; and, in *England*, he points out the particular ſpot, where ſome of the more rare are found: Mr. RAY, however, has remarked, that in this reſpect LOBEL has been inaccurate, or truſted too much to his memory; ſince many have been ſought for in vain, in the ſituations he ſpecified. Frequent reference is made in the margin to the figures in FUCHSIUS,

FUCHSIUS, MATTHIOLUS, DODONÆUS, as far as p. 200; after which, this aſſiſtance is wanting. LOBEL'S own figures are ſmall, and inſufficient in many caſes to expreſs the habit of the plant, the delineation of which, was almoſt the extent of the efforts of thoſe days.

LOBEL having carefully ſtudied the antients, on the *Materia Medica*; having travelled much, and ſeen plants in various countries, was enabled to exerciſe critical ſkill, and to detect numerous errors in the diſpenſation of ſimples, which he does not fail to point out. His ſtrong attachment to the ſtudy intereſted him powerfully in the inveſtigation of new plants, and enabled him to make large acceſſions to knowledge. He travelled over various parts of *England*, and diſcovered many vegetables before unnoticed. He added to the *graſſes* a number of new ſpecies; and, although his ſtile is univerſally condemned as harſh and incorrect, and his deſcriptions frequently obſcure and inſufficient, the *Adverſaria* has, on the whole great merit, abounding with much curious intelligence, and ſome new diſcoveries.

The ſecond part of the *Adverſaria* is but a ſmall part of the whole. It preſents us with a liſt of one hundred and thirty ſpecies of graſſes, known to the author: this is followed by the figures and deſcriptions of ſome new and rare kinds, of the ſame tribe. A number of new plants of the liliaceous and bulbous-rooted order; a copious account, with a figure, of the *yucca*, lately introduced; concluding with a catalogue from CLUSIUS, of thirty-eight varieties of *Anemone*—a proof of the flouriſhing ſtate of the Floriſt's art, in the beginning of the laſt century; at which time it is certain, from LOBEL's book, that many people were very aſſiduous in the cultivation of exotics.

In 1576, LOBEL publiſhed a book, well known, and much quoted ſince, by the name of "OBSERVATIONES; *ſive Stirpium Hiſtoriæ, cui annexum eſt Adverſariorum Volumen. In fol. cum Iconibus.*"

By the aſſiſtance of *Plantin*, this volume was accompanied with 1486 figures, which had been cut for the works of CLUSIUS, MATTHIOLUS, and DODONÆUS.

In 1581 it was tranſlated into Dutch, together

gether with the *Adverſaria*, and the figures augmented to the number of 2116. The ſame year the *icons* were ſeparately caſt off, on paper of the oblong form; the figures amounting to 2191. Some of theſe impreſſions were accompanied with an index, in ſeven languages, which rendered it a very popular book for many years. It preſerves ſome value to this day, as being the edition that LINNÆUS quotes throughout his works.

LOBEL had meditated a very large work, which was to have borne the title of "ILLUSTRATIONES PLANTARUM;" but he lived not to finiſh it. Some of his papers fell into the hands of PARKINSON, and were incorporated into his *Theatrum*. A fragment of the above-mentioned work was publiſhed by Dr. HOW, in 1655; which contains the deſcriptions of many graſſes, and other plants newly diſcovered, or lately introduced. Of the graſſes, many here recorded were firſt diſcovered by LOBEL. The preface contains ſome ſevere cenſures on GERARD, and reflexions on the treatment LOBEL had received from bookſellers; all written

written in a ſtile very reprehenſible in a man of letters. He may be juſtly accuſed of uncandid and diſingenuous conduct towards GERARD, whom, while living, he had treated with the appearance of friendſhip and eſteem, and of whoſe abilities, and zeal, he had ſpoken in the higheſt terms; as is manifeſt in various parts of the *Adverſaria*, in the atteſtation to the catalogue of GERARD's Garden, and by the recommendatory letter prefixed to his Herbal.

I regret that I am not able to do more than barely enumerate the following perſons, who were zealous promoters of gardening, and botanical knowledge, in the time of LOBEL, and liberal in their communications to him.

Dr. *James* CARGIL, of *Aberdeen*; of whom, however, ſome brief mention will be made hereafter.

Edward SAINTLOO, Eſq; of *Somerſetſhire*, whom he ſpeaks of as much attached to ſtudies of this kind.

James COEL, of Highgate, ſon-in-law to LOBEL.

J. NAS-

J. NASMYTH, ſurgeon to James the Firſt.

John De FRANQUEVILLE, a merchant in London; a celebrated floriſt, and a great lover of all rare plants, as well as flowers; from whoſe care, as Parkinſon ſays, "is "ſprung the greateſt ſtore that is now "flouriſhing in this kingdom."

Hugh MORGAN, apothecary to queen *Elizabeth*; of whoſe garden very frequent mention occurs, in both parts of the *Adverſaria*; and alſo in GERARD's Hiſtory afterwards, who ſtiles him "a curious con- "ſervator of ſimples."

William COYS, of *Stubbers*, in the pariſh of *North Okington*, in *Eſſex*, poſſeſſed a garden, which both LOBEL* and GERARD inform us, was richly ſtored with exotics. Under his care, the *yucca* firſt flowered in *England*, in the year 1604.

To

* The name of LOBEL was perpetuated by PLUMIER, who gave it to a plant, which is a native of both the Indies, ſince denominated *Scævola*. But the Swede has preſerved the name to a numerous ſet of plants of the *ſyngeneſious* claſs, among which rank the *cardinal* flowers, and two Engliſh ſpecies.

PLUMIER

To theſe muſt be added the well-known names of GERARD and PARKINSON.

NEWTON.

There is "an Herbal to the Bible," ſaid to be written by *Thomas* NEWTON, and printed in 1587. 8°. This author, after having practiſed phyſic, became a divine and ſchoolmaſter, at *Ilford*, in *Eſſex*; where he died in 1607. His book, I believe, is only a tranſlation of "LEVINI LEMNII *Explicatio Similitudinum quæ in Bibliis ex herbis et arboribus ſumuntur.*" LEMNIUS, who was a phyſician in the province of *Zealand*, briefly deſcribes the plants of the holy Scriptures, and produces a number of curious philological obſervations reſpecting the uſes of plants in ceremonial and ſacred rites. He alſo wrote a memorable work, *De Miraculis occultis Naturæ*. The ſingular pro-

PLUMIER alſo commemorated PENA, by giving his name to one of his new American plants; which, as it proved to be a ſpecies of *Polygala*, was transferred by the author of the ſexual ſyſtem, to an Ethiopian plant of the tetrandrous claſs, though allied in habit to the *Ericæ* and *Paſſerinæ*.

perty

perty of madder in colouring red the bones of animals that are fed with it, appears to have been known to *Lemnius*; but whether he learnt it from *Mizaldus*, or the latter from him, I know not. His book was among the earlieſt productions in its way, and ſeems to have been well received, as may be judged by its paſſing through twelve or thirteen editions, from its firſt publication in 1563 to 1627.

I conceive this *Thomas* NEWTON to have been the writer of thoſe commendatory lines prefixed to LYTE's *Herbal*; in which, after complimenting the author for his judicious ſelection of uſeful knowledge from former writers, he has *verſified*, in leſs than two pages, the names of more than two hundred worthies in medical ſcience, from the earlieſt antiquity to his own times.

CHAP.

CHAP. 9.

Account of Dodoens, *and his* Pemptades, *as introductory to the* Herbal *of* Gerard — *Circumstances of the times favourable to* Gerard.

Account of Gerard—*The catalogue of his garden*—*Account of his* Herbal; *a popular work for more than a century*—*Contemporary Botanists* : Hesketh—Garet : *the correspondent of* Clusius—Lete, *and others.*

GERARD.

LOBEL's writings, howsoever esteemed by the learned, having never been translated into English, could not become popular; and, at the conclusion of the sixteenth century, TURNER's book was, probably no less obsolete, than LYTE's was imperfect. These circumstances, conspiring with the growing taste of the times for gardening, it may be presumed, incited GERARD to undertake his Herbal: a work which maintained its credit and esteem for more than a century; and, pleasing as it is to reflect on the rapid progress and improvement of Botany, within the last half century, yet there are many now living who can

can recollect, that when they were young in ſcience, there was no better ſource of Botanical intelligence, in the *Engliſh* tongue, than the Herbals of GERARD and PARKINSON.

It has been obſerved, that the early edition of DODOENS's book, as tranſlated by CLUSIUS, had been the baſis of LYTE's Herbal; and, as the laſt edition of the ſame author became the foundation of GERARD's, this circumſtance renders it not unſuitable here to take ſome notice of an author, although a foreigner, to whom he owed ſo much of that credit, which has preſerved his memory to the preſent times.

Rembert DODOENS, or DODONÆUS, was born in 1517, near *Mechlin* in *Flanders*. He became conſpicuous for his various erudition when young; was phyſician for ſome time to the Emperor *Maximilian*, and his ſon *Rodolph* II. The importunity of his friends procured his diſmiſſion from the Emperor's ſervice, and he ſettled at *Antwerp*; was afterwards profeſſor at *Leyden*, and died in 1586. He wrote on aſtronomy, geography, and phyſic; but is remembered now,

now, principally, by his botanical works. His attachment to this ſtudy, and the opportunities he enjoyed of gratifying it, enabled him to turn it to the moſt advantageous purpoſes. He began to publiſh in 1552, and continued his acceſſions and improvements to the year 1583, when he collected all his writings, on this ſubject, into one volume, under the following title, "STIRPIUM HISTORIÆ PEMPTADES *Sex, ſive Libri* XXX. *Ant. ex officin. Plant.*" *in folio. cum icon.* 1341. *pp.* 872. Each Pemptade is divided into five books.

The 1ſt comprehends a number of diſſimilar plants in alphabetic order.

2. Flower-garden plants; and the umbelliferous tribe.

3. Medicinal roots: purgative plants: climbing and poiſonous plants: ferns, moſſes, and fungi.

4. Grain: pulſe: graſſes: water and marſh plants.

5. Edible plants: gourd plants: eſculent roots: oleraceous: thiſtles and ſpinoſe plants.

6. Shrubs and trees.

It was reprinted in 1612 and 1616, with ſome

ſome ſmall additions, and being tranſlated alſo into Dutch, with great enlargement, became a popular book in that language.

The judicious ſelection of all that was uſeful, relating to the ſuppoſed plants of the *Materia Medica* of DIOSCORIDES, and of the Arabians, the introduction of all the new ſpecies from CLUSIUS, and other diſcoveries of the time, added to the inſtruction and embelliſhment derived from the figures, which exceeded in number thoſe of any preceding author, rendered *Dodoens*'s book uſeful to the medical profeſſion throughout the world. It ſtill preſerves ſome value, as being referred to by LINNÆUS, for the illuſtration of the European plants.

As GERARD could not attempt an entire new work, there was then extant no other to which he could give the preference, as a baſis to his deſign; for as ſuch only it muſt be conſidered, ſince the interval of time between the publication of DODONÆUS's work in 1583, and the printing of his own "Herbal," had given him opportunities to interſperſe large additions, both in exotic, and indigenous Botany. In this interval

terval the ſcience had been augmented, and not leſs enriched, by the writings of CÆSALPINUS, in 1583; by the *Epitome* of CAMERARIUS, in 1586; by the *Hiſtoria Lugdunenſis* of DALECHAMP, in 1587; by the *Sylva Harcynia* of THALIUS; and eſpecially by the *Hiſtoria* and *Icones* of TABERNÆMONTANUS, in 1588 and 1590.

To theſe may be added, a number of collateral reſources, which the growing commerce and ſpirit of the times rendered favourable to his purpoſes. I will briefly mention the following: the *Materia Medica* had, for a ſeries of years, been perpetually augmenting, by a variety of new drugs, which were eagerly ſought after, the origin of which, notwithſtanding, was in many inſtances obſcure, and in others as yet unknown. At length the publication of GARCIAS *ab* HORTO on the ſimples of the Eaſt Indies, of MONARDES on thoſe of the Weſt, and afterwards of *Chriſtopher à* COSTA's book, ſatisfied, for a time, the impatience of the public.

Theſe authors were tranſlated into Engliſh. *James* FRAMPTON, a merchant of London,

London, who had resided long at *Seville*, from whence he returned in 1576, translated MONARDES into English the next year, under the title of "Joyful News out "of the New Founde World, from the Spa-"nish of *Monardus*," in 4°. CLUSIUS put GARCIAS *ab* HORTO into Latin, in 1567; and *James* GARET had also translated from the Spanish the work of *à* COSTA. These books were incentives to curiosity; and the thousand novelties which were brought into *England* by our circumnavigators, RALEIGH and CAVENDISH, in 1580 and 1588, excited a degree of attention, which at this day cannot, without the aid of considerable recollection, be easily conceived. RALEIGH himself appears to have possessed a larger share of taste for the curious productions of nature, than was common to the seafaring adventurers of that period. And posterity will rank these voyagers among the greatest benefactors to this kingdom, in having been the means, if tradition may be credited, of introducing the most useful root that Providence has held forth for the service of man. A voyage round the globe, howſoever familiarized in ours, was in that age a moſt intereſting and fruitful occaſion of enquiry.

The return of RALEIGH, and the fame of his manifold diſcoveries and collections, brought over from the continent the celebrated CLUSIUS, then in the 55th year of his age. He, who added more to the ſtock of Botany in his day, than all his contemporaries united, viſited ENGLAND, for the third time, to partake, at this critical juncture, in the general gratification.

At this eventful period, GERARD was in the vigour of life, and without doubt felt the influence, and reaped the advantage of all the circumſtances I have enumerated.

John GERARD was born at *Nantwich*, in *Cheſhire*, in the year 1545, and was educated a ſurgeon. He removed to *London*, where he obtained the patronage of the great *Lord Burleigh*, who was himſelf a lover of plants, and had the beſt collection in his garden of any nobleman in the kingdom. GERARD had the ſuperintendance of this fine garden, and retained his employment, as he tells us himſelf, for twenty years.

years. He lived in *Holborn*, where alſo he had a large *phyſic* garden of his own; which was probably the firſt of the kind in *England*, for the number and variety of its productions. It ſhould ſeem, that in his younger days he had taken a voyage into the Baltic, ſince he mentions having ſeen the wild pines growing about *Narva*.

GERARD appears alſo to have been favoured by the college of phyſicians, and is highly extolled by Dr. BULLEYN. Both LOBEL, and Dr. BROWNE, phyſician to the queen, wrote, in Latin, commendatory letters to him, on the publication of his Herbal. He attained to ſuch eminence in his profeſſion, as to be choſen maſter of the company. He died about the year 1607.

There is a half ſheet print of GERARD prefixed to his own edition of the "Herbal," done in the 53d year of his age, and a ſmall oval one at the bottom of a full half ſheet frontiſpiece, before JOHNSON's edition.

The earlieſt publication of GERARD was the liſt of his own garden in Holborn, under the following title, " *Catalogus Arborum, Fruticum, ac Plantarum, tam indigenarum quam exoticarum, in horto* JOHANNIS GERARDI, *civis ac chirurgi Londinensis nascentium.* Impensis J. Norton, 1596." 4°. and again in 1599.

The first edition was dedicated to Lord BURLEIGH; but that nobleman dying before the publication of the second, it was inscribed to his patron, *Sir Walter* RALEIGH.

This little piece, from the nature of the publication, is become very scarce. I believe there is only a manuscript copy of it in the collection of *Sir* JOSEPH BANKS.

We are informed, in the life of Dr. BULLEYN, that GERARD's Garden contained near eleven hundred sorts of plants, of foreign and domestic growth; from whence, says Mr. *Oldys*, " it may appear, that our " ground would produce other fruits be- " sides hips and haws, acorns and pignuts;" for at this time, " kitchen-garden wares " were imported from Holland, and fruits " from France." There are one thousand and

and thirty-three ſpecies in this Catalogue, and the following atteſtation, written by LOBEL, is annexed.

"*Herbas, ſtirpes, frutices, ſuffrutices, et arbuſculas hoc catalogo recenſitas, quamplurimas ac fere omnes me vidiſſe Londini in horto Johanni* GERARDI, *chirurgi et botanici peroptimi (non enim omnes eodem ſed variis temporibus anni pullulaſcunt, enaſcuntur et florent). Atteſtor Matthias De* LOBELL, *ipſis calendis Junii* 1596."

In 1597, came out his "HERBAL, or "GENERAL HISTORY OF PLANTS;" printed by John Norton, in folio; and ſome authors mention another impreſſion in 1599.

That the foundation of this work was a tranſlation of DODOENS's Herbal, a compariſon of the two aſcertains beyond a doubt. LOBEL, both in his animadverſions on RONDELETIUS, and in his *Stirpium Illuſtrationes*, informs us, that Dr. PRIEST, at the expence of Mr. *Norton*, had been engaged to make a tranſlation of DODONÆUS's *Pemptades*; and, dying ſoon after he had finiſhed it, the manuſcript came into GERARD's

 hands;

hands; who has been cenſured for having endeavoured to conceal his poſſeſſing theſe papers, and for aſſuming to himſelf the merit of the tranſlation, when it is generally agreed, that his knowledge of the Latin language was not equal to ſuch an undertaking. LOBEL, indeed, judged the ſame of Dr. PRIEST, and points out inſtances of his inſufficiency. It muſt, however, be allowed, that GERARD is not backward in confeſſing his want of ſkill in the learned languages. LOBEL farther informs us, that when the work was in the preſs, and that part of the firſt book printed relating to graſſes, his friend, *James* GARET, a perſon eminently ſkilled in flowers and exotics, admoniſhed *Norton* of ſome groſs errors; on which, the printer engaged LOBEL to ſuperintend the work; that he actually did correct it "in a thouſand places;" and that there were many other miſtakes, which GERARD would not allow him to alter, alleging that it was ſufficiently correct, and that "LOBEL had forgotten the Engliſh "language."

In order further to conceal his plagiariſm, LOBEL

LOBEL adds, that he has inverted the diſtribution of the chapters in DODOENS's book, and adopted that of the *Adverſaria*. This may be conſidered as a futile objection, and even turned into an approbation of LOBEL's method; but he charges him alſo with largely plundering the *Adverſaria*, without any acknowledgment.

GERARD compriſes the whole vegetable kingdom in three books. The *firſt* contains the graſſes, grain, ruſhes, reeds, flags, and bulbous-rooted plants. The *ſecond*, all herbs uſed in diet, phyſic, or for ornament and pleaſure. The *third*, trees, ſhrubs, fruit-bearing plants, roſins, gums, roſes, heaths, moſſes, muſhrooms, and ſea plants. The whole divided into upwards of eight hundred chapters, which, in the arrangement of that time, may, if the expreſſion is allowable, be conſidered as ſo many genera.

In each chapter the ſeveral ſpecies are deſcribed; then follow the place, time of flowering, names, and virtues.

The figures Mr. *Norton* procured from *Frankfort*, being the ſame blocks which had been uſed for the Dutch Herbal of TABER-

NÆMONTANUS

NÆMONTANUS in 1588. In this manner, GERARD, with DODOENS for his foundation, by taking in alſo many plants from CLUSIUS, and from LOBEL, by the addition of ſome from his own ſtock, publiſhed a volume, which, from its being well timed, from its comprehending almoſt the whole of the ſubjects then known, by being written in Engliſh, and ornamented with a more numerous ſet of figures than had ever accompanied any work of the kind in this kingdom, obtained great repute. To this we muſt add the fortunate circumſtance of its acquiring afterwards ſo learned an editor as JOHNSON, which eſtabliſhed the character of it, and gave it precedence as a popular book, for more than a century. And notwithſtanding his manifeſt inferiority to LOBEL in point of learning, it muſt yet be owned, that GERARD contributed greatly to bring forward the knowledge of plants in *England*. His connection with the great, and his ſituation in *London*, favoured an extenſive correſpondence, both with foreigners and his own countrymen; and his ſucceſs in procuring new exotics, as well as ſcarce indigenous

indigenous plants, was equal to his diligence and aſſiduity. In fact, we owe to GERARD and his friends the diſcovery of ma y new Engliſh plants; and his name will be remembered by botaniſts with eſteem, when the utility of his Herbal is ſuperſeded. That he was conſidered as poſſeſſing a very extenſive ſhare of this ſcience, we are juſtified in believing, on the teſtimony of Mr. *George* BAKER, chief ſurgeon to the queen, who aſſures us, that he ſaw him "tried with one of the beſt ſtran ers that "ever came into *England*, and was ac-"counted in *Paris* the only man, being "recommended to me," ſays BAKER, "by "that famous man, AMBROSE PAREY; "and he being here, was deſirous to go "abroad with ſome of our herbariſts, for "the which I was the mean to bring them "together, and one whole day we ſpent "therein, ſearching the rareſt ſimples: but "when it came to the trial, my French-"man did not know one to his four *."

* PLUMIER gave the name GERARDIA to a plant of the *didynamous* claſs, diſcovered in the tropical regions of America; to which LINNÆUS has ſince added five ſpecies.

Among

Among the many who promoted GERARD's work by their communications, I muſt not omit the names of *Thomas* HESKETH, of *Lancaſhire*; *Thomas* EDWARDS, apothecary, at *Exeter*; both ſkilled in the knowledge of Engliſh plants.

James GARET, of *London*, apothecary, " a curious ſearcher of ſimples." He was the correſpondent of CLUSIUS, to whom he communicated a great number of natural curioſities, particularly of exotic growth, and is mentioned with great reſpect by that learned foreigner, in numerous places of his *Libri Exoticorum*. He ſeems to have been one of the principal cultivators of tulips, which he propagated by ſeeds and bulbs for twenty years, every ſeaſon bringing forth, as GERARD obſerves, " new plants of ſundry colours not before ſeen, all which to " deſcribe particularly, were to roll *Siſiphus*'s ſtone, or number the ſands."

I find three perſons of the ſame name, *James* GARET the father, and *James* the ſon, and *Peter*, as I ſuppoſe, the brother of *James* the elder. PARKINSON, ſpeaking probably of the laſt, informs us, that he was originally a druggiſt in Lime-ſtreet.

† He

He was, I believe, the tranſlator of *à* COSTA, as hath been before noted.

Mr. Bredwell, " practitioner in phyſic, a " learned and diligent ſearcher of ſimples," in the weſt of *England*.

Mr. *Nicholas* LETE, a merchant of London, " greatly in love with rare and faire " flowers, for which he doth carefully ſend " into *Syria*, having a ſervant there at " *Aleppo*, and in many other countries; for " which myſelf and the whole land are " much bound unto him."

Dr. *John* MERSHE, of *Cambridge*.

Mr. *James* COLE, a merchant of *London*, " a lover of plants, and very ſkilful in the " knowledge of them."

Among thoſe of eminent ſtation, who patroniſed the ſcience, GERARD does due honour to Sir *Walter* RALEIGH; Lord *Edward* ZOUCH, the patron of LOBEL, who brought plants and ſeeds with him from *Conſtantinople*; and to Lord HUNSDON, Lord High Chamberlain of *England*, who, he ſays, " is worthy of triple honour for " his care in getting, as alſo for his curi- " ous keeping, ſuch rare and ſtrange things " from the fartheſt parts of the world."

CHAP. 10.

Johnſon *the improver of* Gerard's *book—Anecdotes of—His* Iter in Agrum Cantianum *the firſt* Engliſh *local catalogue—Enters into the king's army, and is killed at the ſiege of* Baſing—*His edition of* Gerard—Mercurius Botanicus—*Verſion of* Parey's *works.*

Contemporary aſſiſtants — Goodyer — Bowles — Tunſtal — Glyn — Morgan.

JOHNSON.

THOMAS JOHNSON was born at *Selby*, in *Yorkſhire*, and bred an apothecary in *London*. He afterwards kept a ſhop on *Snow-Hill*, "where, by his unwearied pains, advanced with good natural parts," ſays Mr. *Wood*, "he attained to be the beſt herbaliſt of his age in *England*."

He was firſt announced to the public, by a ſmall piece under the title of "ITER IN AGRUM CANTIANUM, 1629; *et* ERICETUM HAMSTEDIANUM, 1632: which were

were the firſt local catalogues publiſhed in *England*. He ſoon after acquired great credit by his new edition and emendation of GERARD's "Herbal."

In the civil wars, his zeal for the royal cauſe led him into the army, in which he greatly diſtinguiſhed himſelf; and the univerſity of *Oxford*, in conſideration of his merit and learning, added to that of his loyalty, conferred upon him the degree of doctor of phyſic, May 9, 1643.

In the army, he had the rank of lieutenant colonel to Sir *Marmaduke* RAWDON, governor of *Baſinghouſe*. Mr. *Granger* informs us, that "he ſet fire to the *Grange*, "near that fortreſs, which conſiſted of "twenty houſes, and killed and burnt about "three hundred of Sir William *Waller's* "men, wounded five hundred more, and "took arms, ammunition, and proviſions "from the enemy." Wood adds, "that "going with a party on the 14th of September, 1644, to ſuccour certain of the "forces belonging to that houſe, which "went to the town of *Baſing* to fetch pro-"viſions thence, but beaten back by the

" enemy, headed by that notorious rebel,
" Colonel *Richard Norton*, he received a
" ſhot in the ſhoulder, of which he died in
" a fortnight after. At which time his
" worth did juſtly challenge funeral tears;
" being then no leſs eminent in the garri-
" ſon for his valour and conduct as a ſol-
" dier, than famous through the kingdom
" for his excellency as an herbaliſt and
" phyſician."

I have mentioned Johnſon's *Iter Cantianum*, and *Ericetum Hamſtedianum*; but not having ſeen either, I can give no account of them.

In 1633, he publiſhed his improved edition of GERARD, under the title of " The
" HERBAL, *or* GENERAL HISTORY *of*
" PLANTS, gathered by *John* GERARD,
" of *London*, very much enlarged and a-
" mended by *Thomas* JOHNSON, citizen and
" apothecary of *London*, for Iſlip and Nor-
" ton." 1633. fol.; and again 1636. pp. 1630.

An interval of thirty-ſix years, from the date of *Gerard's* work, had effected a great change in the ſtate of botanical knowledge; many

many new plants had been introduced, and many valuable works publiſhed on the continent, particularly the *Hortus Eyſtettenſis* in 1613, and the *Prodromus* of Bauhine in 1620. No publications had appeared at home, except ſuch as were adapted to the Floriſt and Gardener; *Gaſpar Bauhine*'s invaluable *Pinax* had facilitated and ſhortened the labour of conſulting preceding authors. All theſe circumſtances were favourable to JOHNSON; and his acknowledged ſuperiority to GERARD in the learned languages, might juſtly raiſe the expectation of the public; inſomuch that it becomes a matter of ſpeculation, why JOHNSON acquieſced in the character of an editor only. It may indeed be converted into a ſtrong preſumption of the value ſet by the public on GERARD; which probably prevented the riſk of a new title.

The general expectation was not diſappointed. The advantages above noted enabled JOHNSON to amplify and improve his author to ſuch a degree, that his book eminently deſerves the *encomium* that HALLER has beſtowed upon it, when he calls it "*dignum opus, et totius rei herbariæ eo* "*ævo notæ, compendium.*"

After what has been ſaid of the plan, as it ſtands in GERARD, it remains only to ſhew briefly what JOHNSON has done. In about twelve pages, he has prefixed a conciſe, candid, and judicious account of the moſt material writers on the ſubject, from the earlieſt ages to the time in which he wrote; concluding with a particular account of his own work, from its origin in Dr. PRIEST's tranſlation. After this follows a table, pointing out, with great preciſion, all his additions; by which we learn, that he enriched the work with more than eight hundred plants not in GERARD, and upwards of ſeven hundred figures, beſides innumerable corrections. By procuring the ſame cuts that GERARD uſed, (to which collection a conſiderable acceſſion had been made) and by having ſome new blocks cut, his work contained a greater number of figures than any Herbal extant; the whole amounting to 2717. He informs us, in an apology he makes for not inſerting his additional matter in the edition of 1636, that

that he intended to travel throughout the kingdom in ſearch of the more rare plants, and afterwards to compriſe all his diſcoveries in an appendix.

In 1634, he publiſhed " MERCURIUS BOTANICUS; *ſive* PLANTARUM *gratia ſuſcepti Itineris, anno* 1634, DESCRIPTIO; *cum earum Nominibus Latinis et Anglicis*." *Lond.* 8vo. pp. 78.

It is dedicated to *Sir Theodore* MAYERNE, and others of the college, in his own, and the names of his aſſociates in the excurſion, who were all of the company of the Apothecaries. It was the reſult of a journey, through *Oxford*, to *Bath* and *Briſtol*, and back by *Southampton*, the Iſle of *Wight*, and *Guildford*, made with the profeſſed deſign to inveſtigate rare plants. He has deſcribed, in not inelegant Latin, their rout, which took up only twelve days, and the agreeable reception they met with among their medical acquaintance. We meet with a liſt of exotics, amounting to 117, cultivated by Mr. *George* GIBBS, a ſurgeon at *Bath*, who had made a voyage to *Virginia*, from whence he brought many new plants; which, as it exhibits the advanced ſtate of

K 2 gardening

gardening in this country at that time, is now a matter of curioſity.

The plants of ſpontaneous growth enumerated in this ſhort tour, varieties being excluded, exceed ſix hundred, which, at a time when the *cryptogamiæ* were ſcarcely noticed, and in the ſeaſon when neither the very early nor late plants could be ſeen, is no inconſiderable number. In this catalogue are ſeveral not diſcovered in *England* before. With this tour JOHNSON gave his ſmall tract, "*De Thermis Bathonicis, ſive earum deſcriptio, vires, utendi tempus, modus*, &c." Lond. 1634. pp. 19. There are three ſmall plans of the baths, and one of the city, which ſeem to be copied from Speed's map. Theſe are now pleaſing curioſities to the lovers of antiquity, and to all who contemplate the aſtoniſhing increaſe of the city ſince that time.

This was followed by "PARS ALTERA, *ſive* PLANTARUM *gratia ſuſcepti Itineris in Cambriam ſeu Walliam* DESCRIPTIO." Lond. 1641. 8°.

JOHNSON, if not the firſt, was among the earlieſt Botaniſts who viſited *Wales*, and *Snowdon*, with the ſole intention of

diſcovering

difcovering the rarities of that country in the vegetable kingdom. The journey feems to have anfwered his purpofe, and afforded him a rich harveft. In this expedition he firft found the yellow poppy, *papaver cambricum:* mountain faw-wort, *ferratula alpina:* rofe-root, *rhodiola rofea*; and feveral other plants.

I cannot afcertain the age of JOHNSON at his death, but there is reafon to think he could not be far advanced in life, if indeed he was arrived at the meridian of it. I ground my opinion on the circumftance of LOBEL's total filence relating to him, in his *Adverfaria*, printed in 1605. Engaged as JOHNSON was, in the exercife of a profeffion, which, independent of the calls of duty, demands much facrifice of time, to the forms and civilities of life, his HERBAL is an ample teftimony of zeal and induftry. I do not find that he was the author of any other publications, than thofe, of which I have given fome account; but, he tranflated the works of *Ambrofe* PAREY, which he publifhed at *London* in 1643. They were reprinted, if I miftake not, for the laſt time in 1678. This excellent man, who in the character of ſurgeon, ſucceſſively ſerved four ſovereigns of *France*, was attached to the proteſtant cauſe; and for his extraordinary merit, and his having cured *Charles* IX. of a tendon wounded in bleeding, was ſaved from the maſſacre of St. *Bartholomew*. He ſurvived this event 19 years, and died in 1590. His works were collected by himſelf, in 1582, in folio, and ran through nine or ten editions on the continent. PAREY's improvements in his profeſſion had been ſingularly important; there can be no doubt, therefore, that our author performed a very acceptable ſervice to his countrymen, by putting his writings into an Engliſh dreſs *.

* MILLER conſecrated the name of JOHNSON by aſſigning it to a berry-bearing ſhrub of *Carolina*, belonging to the *tetrandrous* claſs; firſt figured by PLUKENET, *tab.* 136. f. 3. and ſince by CATESBY, *vol.* 2. *tab.* 47. The Engliſh Botaniſts, who muſt conſider JOHNSON as entitled to ſo honourable a diſtinction among their worthies, will regret that his name ſhould not be retained in the Linnæan ſyſtem, in preference to *Callicarpa*, by which term this ſhrub is now well known in the Engliſh gardens.

Before

Before JOHNSON is dismissed, it would be unjust not to notice some of those, to whom the author was especially indebted for assistance, and for the communication of English plants. Among these, the first place is due to Mr. *John* GOODYER, of *Maple Durham*, in *Hampshire*, whose name occurs repeatedly in GERARD's "Herbal," and very frequently in PARKINSON's, in which he is stiled "a great lover and "curious searcher of plants; who, besides "this" (speaking of the *geranium saxatile*) "hath found in our country many "other plants, not imagined to grow in "our land." He seems not only to have been what may be called a *practical* Botanist, but learned, and critically versed in the history of the science. This may be fairly inferred from his curious communication, relating to the manuscripts under the name of APULEIUS *Madaurensis*, and from his observations on the *saxifrage* of the ancients, inserted at p. 604. The great number of rare English plants, which Mr. GOODYER first brought to light, entitles him to the moſt reputable rank among thoſe who have advanced the botanical knowledge of this kingdom.

Mr. *George* BOWLES, of *Chiſſelburſt*, in *Kent*, alſo diſtinguiſhed himſelf by his ſucceſsful inveſtigation of many new plants. He ſpent ſome time in *Wales*, where his diſcoveries were very ample; and he is mentioned with particular attention, in numerous inſtances, by our author.

The names of JOHNSON's aſſociates in his Kentiſh, and other ſimpling excurſions, occur in the preface; and in the body of the work we meet alſo with the following:

John TRADESCANT the elder, who became famous afterwards for his fine garden, and muſeum of natural curioſities.

Sir *John* TUNSTAL, gentleman uſher to the queen, is recorded as poſſeſſing a garden at *Edgcome* in Surrey, ſtored with plants, which are ſaid to have belonged to the queen.

Mr. *Thomas* GLYN, who firſt found that elegant plant the *gnaphalium marinum*, on the coaſt of Wales.

Mr.

Mr. *Hugh* MORGAN, apothecary to the queen, before mentioned under the article of LOBEL.

Mr. *Robert* ABBOT, of *Hatfield*, near St. *Albans*, a learned preacher, and an excellent and diligent herbarist.

BOELIUS or BOEL, of whom further notice more properly comes under the article of PARKINSON.

Mr. *John* REDMAN, " a skilful herba-" rist," an inhabitant of the northern part of England.

Frequent and respectable notice is also taken of Mr. *John* PARKINSON, the subject of the succeeding article. His *Paradisus Terrestris* is much commended, and his garden referred to as abounding in choice plants.

CHAP. II.

Parkinſon—*Brief account of his life—His* Paradiſus : *the beſt view of the ſtate of the flower garden in that age*—Theatrum Botanicum : *a more original and laboured performance than* Gerard's Herbal—*Its merit not ſufficiently acknowledged by his ſucceſſors.*

Boel : *and other contemporaries of* Parkinſon.

PARKINSON.

JOHN PARKINSON was born in 1567. I regret that I am not enabled to ſupply a more ample account of this laborious man, whoſe learning and abilities appear to me not to have been juſtly appreciated. He was bred an apothecary, and lived in *London*. He was contemporary with GERARD and LOBEL, during the latter part of their lives; and ſurvived JOHNSON ſeveral years. LOBEL, in the ſecond part of his *Adverſaria*, and JOHNSON, in his *Gerardus Emaculatus*, ſpeak of him as a man of eminence in his profeſſion, and as poſſeſſed of a garden well ſtored with rarities. In fact, he roſe to ſuch a degree of reputation as to be appointed apothecary to King *James*; and at the publication of his "Theatre of "Plants," he obtained, as we learn by Sir *Theodore* MAYERNE's commendatory letter prefixed to it, the title from *Charles* the Firſt of *Botanicus Regius Primarius*. The time of his death I cannot aſcertain; but, as his "Herbal" was publiſhed in 1640, and he appears to be living at that time, he muſt have attained his 73d year. There is a print of him prefixed to his *Paradiſus*, in the 62d year of his age, and a ſmall oval one, in the title of his "Herbal," or "Theatre of Plants."

His firſt publication was the "PARA-"DISI IN SOLE PARADISUS TERRESTRIS; "or, a garden of all ſorts of pleaſant flowers, "which our Engliſh ayre will permit to be "nurſed up: with a kitchen garden of all "manner of herbs, roots, and fruits, for "meat or ſauſe, uſed with us, and an or-"chard of all ſorte of fruit-bearing trees and "ſhrubbes fit for our land; together with "the right ordering, planting, and preſerv-"ing

" ing of them, and their uſes and vertues,
" Collected by *John* PARKINSON, apothe-
" cary of London 1629." Folio. pp. 612.

There was a ſecond edition publiſhed after the author's death, corrected and enlarged, in 1656.

As the ſubject of this book intereſts the floriſt and gardener merely, it comes leſs within the ſcope of this work than his " Herbal." It is dedicated to Queen *Elizabeth*; and, agreeably to the panegyrical cuſtom of the times, is ſet off with recommendatory verſes; among which we meet with ſome in Latin from *Thomas* JOHNSON, doubtleſs the editor of GERARD, and a Latin letter, in a high ſtrain of eulogy, from Sir *Theodore* MAYERNE.

The plants are arranged without any other order than that expreſſed in the title page. Garden flowers are divided into 134 chapters, according to the generical names of the time; kitchen plants into 63 chapters; fruit trees and ſhrubs into 24 chapters; and a corollary of 22 ſpecies. Nearly one thouſand plants are ſeparately deſcribed; of which ſeven hundred and eighty are figur-

ed

ed on one hundred and nine tables, which appear to have been cut on purpofe for this work. Many are copied from CLUSIUS and LOBEL. The figures are lefs commendable for the defign than the execution, and are much inferior, on the whole, to thofe of GERARD'S "Herbal." In the Latin names, the author has made ufe principally of *Cafper* BAUHINE; fome are taken from LOBEL. The mode of arrangement in each chapter is fimilar to that of GERARD. After the defcription of all the fpecies, follow the place, time of flowering, fynonyms, and virtues. Lefs is fpoken of the culture than feems to be requifite.

Several Englifhmen had written on gardening and agriculture in the fixteenth century, of whom the firft on hufbandry, as far as I can find, was *Antony* FITZHERBERT, a famous lawyer and juftice of the King's Bench, whofe "Booke of Huf-"bandrie" was printed firft in 1534. One of the earlieft, if not the firft on gardening, is *Thomas* HILL, "His profytable Art of "Gardening," printed in 1574. The next was, "The new Orchard and Garden," by William

William LAWSON, in 1597. In 1600, Sir *Hugh* PLATT, the author of many other useful tracts, put forth his "Garden of "Eden;" a book of great merit in its time. All these passed through numerous editions, and the last preserved credit to the end of the century.

PARKINSON however, as I apprehend, was the first author, who separately described and figured the subjects of the *Flower Garden*. The *Paradisus Terrestris* is therefore, at this time, a valuable curiosity, as exhibiting the most compleat view of the extent of the English garden at the beginning of the last century. Intertropical productions had been but sparingly imported. The real stove plants are very rare throughout the book. There are some American species, and particularly from *Virginia*, as being a part of that continent with which *England* had the most frequént intercourse. But the principal productions of the English gardens were exotic European, and Grecian plants, some Asiatic, and a few from the northern coasts of *Africa*.

A modern florist, wholly unacquainted with

with the ſtate of the art at the time PARKINSON wrote, would perhaps be ſurprized to find that his predeceſſors could enumerate, beſides ſixteen deſcribed as diſtinct ſpecies, one hundred and twenty varieties of the *tulip*, ſixty *anemonies*, more than ninety of the *narciſſus* tribe, fifty *hyacinths*, fifty *carnations*, twenty *pinks*, thirty *crocuſes*, and above forty of the *iris* genus. In the orchard we find above ſixty kinds of *plums*, as many *apples* and *pears*, thirty *cherries*, and more than twenty *peaches*.

In 1640, PARKINSON publiſhed his "THEATRUM BOTANICUM; or, Theatre "of Plants, or an Herbal of a large extent: "containing therein a more ample and ex- "act hiſtory and declaration of the phyſical "herbs and plants that are in other au- "thors; encreaſed by the acceſs of many "hundreds of new, rare, and ſtrange "plants from all the parts of the world; "with ſundry gummes, and other phyſical "materials, than hath been hitherto pub- "liſhed by any before: and a moſt large "demonſtration of their nature and virtues. "Shewing withal, the many errors, differ- "ences,

" ences, and overſights of ſundry authors
" that have formerly written of them, and
" a certain confidence, or moſt probable
" conjecture of the true and genuine herbs
" and plants: diſtributed into ſundry claſſes
" or tribes, for the more eaſy knowledge
" of the many herbs of one nature and
" property, with the chief notes of Dr.
" *Lobel*, Dr. *Bonham*, and others, inſerted
" therein." London. Folio. pp. 1746.
SEGUIER mentions an edition in 1656,
which I never ſaw, and ſuſpect it was not
a new impreſſion.

This work was the labour of PARKINSON's life, and was not publiſhed until he was arrived at a very advanced period. He tells us, in the preface, that, owing " to the diſ-" aſtrous times," and other impediments, the printing of it was long retarded. Originally it was intended to have contained only the medicinal herbs, under the title of " A phyſical Garden of Simples," but he enlarged his plan, and endeavoured to comprehend all the Botany of his time. It is manifeſt, even from a curſory view of it, that it is a work of much more originality than

than that of GERARD; and it contains abundantly more matter than the laſt edition of that author, with all JOHNSON's augmentations. In the general diſpoſition of the ſubject, the order is chiefly founded on the known, or ſuppoſed qualities, and virtues of the plants; being divided into ſeventeen tribes, as follow:

1. *Plantæ odoratæ.* Sweet-ſmelling plants.
2. *Catharticæ.* Purging plants.
3. *Venenatæ, narcoticæ, nocivæ, et alexipharmicæ.* Venemous, ſleepy, and hurtful plants, and their counterpoiſons.
4. *Saxifragæ.* Saxifrages, or break-ſtone plants.
5. *Vulnerariæ.* Wound herbs.
6. *Refrigerantes, et intubaceæ.* Cooling, and ſuccory-like herbes.
7. *Calidæ, et acres.* Hot, and ſharp-biting plants.
8. *Umbelliferæ.* Umbelliferous.
9. *Cardui, et ſpinoſæ.* Thiſtles, and thorny plants.

10. *Filices, et herbæ capillares.* Ferns, and capillary herbes.
11. *Legumina.* Pulſes.
12. *Cerealia.* Corn.
13. *Gramina, junci, arundines.* Graſſes, ruſhes, and reeds.
14. *Paludoſæ, aquaticæ, marinæ, muſci, et fungi.* Marſh, water, and ſea plants, moſſes, and muſhrooms.
15. *Miſcellaneæ.* The unordered tribe.
16. *Arbores, et frutices.* Trees, and ſhrubbes.
17. *Exoticæ, et peregrinæ.* Outlandiſh plants.
18. *Appendix.*

This heterogeneous claſſification, which ſeems to be founded on that of *Dodoens,* ſometimes on the medicinal qualities, ſometimes on the habit, and on the place of growth, ſhews the ſmall advances that had been made towards any truly ſcientific diſtribution. On the contrary, both GERARD, JOHNSON, and PARKINSON, had rather gone back, by not ſufficiently purſuing the example of LOBEL.

In

In the particular diſpoſition of the ſubjects, under each chapter or genus, PARKINSON follows the rules of GERARD, and JOHNSON, by giving, after the Latin and Engliſh name, the deſcriptions at large; then the place of growth, and time of flowering; the ſynonyms, and laſtly, the virtues and uſes.

Nice diſcrimination of ſpecies from each other, or from varieties, muſt not be expected in this work, more than in GERARD, or his *Emaculator*. Almoſt every Botaniſt was then a Floriſt too. CLUSIUS himſelf, who had enlarged the ſeience, by his own diſcoveries, beyond any other man, continued to raiſe tulips from ſeed, for more than 35 years. PARKINSON's "*Paradiſus*" proves his attachment to the Flower Garden, in the early part of his life; and this bias influenced him throughout the "Theatre of Plants." As yet, no line had been drawn with ſufficient accuracy, between ſpecies and variety, between nature and the effect of culture, or of ſoil and ſituation, nor was this brought about till the eſſential parts of vegetables, the flower, and the fruit, became objects of classification, instead of the vague distinctions hitherto observed; of which it may be sufficient to adduce one example, out of hundreds equally futile. The sea cabbage, (*brassica orientalis*) a *siliquose* plant, is ranked by GERARD and JOHNSON, as well as by PARKINSON, even contrary to the examples of CLUSIUS and DODONÆUS, under the same generical name with the thorow wax, (*bupleurum*) an *umbelliferous* plant, merely because the leaf is of the *perfoliate* kind.

These are defects common to the age, and PARKINSON must not be appreciated by modern improvement, but by comparison with his contemporaries. In this view, if I am not mistaken, he will appear more of an original author than GERARD, or JOHNSON, independent of the advantages he might derive from being posterior to them. His "*Theatre*" was carried on thro' a long series of years, and he profited by the works of some late authors, which, though equally in JOHNSON's power, he had neglected to use. PARKINSON's descriptions, in many instances, appear to be new. He is more

more particular in pointing out the places of growth. In the enumeration of the ſynonyms, he has not only given nearly the whole of BAUHINE's "*Pinax*," but, very frequently, has himſelf conſulted the original authors, and enters minutely into a diſcuſſion of their doubts. In the account of the virtues, and uſes, PARKINSON is diffuſe. It was his profeſſed deſign to make his work a *Materia Medica*; and if, in him, we meet with the qualities of plants eſtimated on Galenical principles, by the degrees of hot and cold, moiſt and dry, &c. it was the theory of the day, from which authors of higher eminence were not emancipated. He not only gives the opinions of the Greek and Roman phyſicians, but of the Arabians, and has tranſlated from the moderns, and his contemporaries, whatever could illuſtrate his ſubject, and render it as perfect as the intelligence of the times would allow. To this end he has extracted largely from CLUSIUS's "Exotics," from D'ACOSTA, MONARDES, and GARCIAS *ab* HORTO on the drugs and ſimples of the Eaſt and Weſt Indies; of which, at that time, many were newly introduced, and imperfectly known.

PARKINSON's work is much more extensive than JOHNSON's, in the number of subjects described, he having taken, as before observed, advantages which the *Emaculator* of GERARD neglected. Many of the plants of *Ægypt*, from *Prosper* ALPINUS, many of the North American, or Canadian plants, from CORNUTUS, and some from COLUMNA's work, are introduced. He neglected no opportunities of procuring new plants from abroad. The nature of his profession did not allow him to make distant or frequent excursions in *England*; but, by the assistance of his correspondents, and some of LOBEL's posthumous writings, which he purchased, he was enabled to enlarge, not only the catalogue of British plants, but to introduce many exotics before unknown.

JOHNSON had described about 2850 plants, PARKINSON has near 3800. These accumulations rendered the "THEATRUM BOTANICUM" the most copious book on the subject in the English language;

guage; and it may be preſumed, that it gained equally the approbation of medical people, and of all thoſe who were curious and inquiſitive in this kind of knowledge. Both this work, and GERARD's afterwards, acquired conſequence by the references of Mr. RAY, who may be ſaid, in the language of the *Catalogus Oxonienſis*, to have raiſed them to claſſical eminence in Engliſh Botany, and preſerved them from oblivion as long as his own works remain. Without any deſign of depriving JOHNSON of his due praiſe, yet it is obvious, from the recollection of certain circumſtances, that PARKINSON laboured under diſadvantages and impediments, which probably tended to depreſs his work at the time, although it had undoubtedly been carrying on through a longer ſeries of years than *Johnſon*'s, and was more copious in its deſign.

JOHNSON had the opportunity that GERARD himſelf obtained, of procuring all the cuts from abroad. PARKINSON's, on the other hand, though copied from the ſame figures, appear to have been cut anew, purpoſely for his work. The delay occaſioned by this circumſtance, beſides the great expence, was, probably, among the obſtacles the author complains of, which ſo long retarded the publication of his work. Add to this, that the figures were after all inferior to the old tables, both in number and execution. JOHNSON's exceed thoſe of PARKINSON, by more than an hundred. Both theſe works may be conſidered as *Digeſts* of the Botany of the age, in the Engliſh tongue; but it is to be feared the ſame cenſure lies againſt them which *Caſpar* BAUHINE lodged againſt DALECHAMP's hiſtory, publiſhed in 1588, in which he demonſtrated, that more than 400 plants were twice deſcribed.

Nor is it wonderful that the attempt to comprehend, and diſcriminate the whole vegetable kingdom, was a plan too extenſive for one man, eſpecially in the augmented ſtate in which PARKINSON found it. The magnitude of the deſign neceſſarily involved a multitude of errors, and expoſed both GERARD and PARKINSON to the cenſures of malignant critics. Had the candour of LOBEL been equal to his learning

learning and knowledge, he had ſpared much of his acrimony againſt theſe induſtrious writers, whoſe laudable endeavours rather merited his applauſe.

Among thoſe contemporaries, whoſe collateral aſſiſtance is acknowledged by PARKINSON, Mr. (or, as he is ſtiled in ſome parts of the work, *Dr.) William* BOEL claims particular notice. He was a native of the Low Countries, and had travelled into various parts of *Germany* and *Spain*; had been in *Barbary*, reſided at *Tunis*, and, at the publication of "the Herbal," lived at *Liſbon*. From all theſe countries he ſent ſeeds of many plants before unknown in *England*. He was the correſpondent of CLUSIUS, and ſeems to have been very zealous for the improvement of natural knowledge.

Mr. *John* GORDIER, "a great lover "and curious ſearcher of plants, who, be-"ſides this," (ſpeaking of the *Geranium lucidum*) "hath found in our country other "plants, not imagined to grow in our "land.

In PARKINSON's works we alſo find the name

name of Mrs. *Thomazin* TUNSTAL, a lady whom he celebrates, not only for her taste in cultivating a garden which was well stored with exotics, but for her knowledge of *English* botany, and her discoveries of several curious vegetables found about *Ingleborough Hill*, in *Lancashire*; which were not known before to grow in *England*. Whether she was allied to Sir *John* TUNSTAL, noticed in the account of JOHNSON, I cannot ascertain.

Besides the names of BOWLES, GOODYER, TRADESCANT, and others, mentioned by JOHNSON, we meet with the following, as having contributed to the general stock. *John* NEWTON, surgeon, at *Colliton*, *Somersetshire*; Dr. *Antony* SADLER, physician at *Exeter*; Mr. *William* QUICK, apothecary, *London*; Mr. BRADSHAUGH, of *Yorkshire*; Mr. SILLIARD, of *Dublin*, and divers others*.

* PARKINSON is commemorated for his botanical labours by PLUMIER, in having his name applied to a *decandrous* tree, a native of the *Caribbee* islands, and of the adjacent continent, well known in the English stoves, and called in Jamaica the *Jerusalem thorn*.

CHAP.

CHAP. 12.

History of wooden cuts of plants—Plantin's *accumulation of these figures*—*Fate of* Gesner's *excellent engravings*—*Of those to the Herbals of* Turner, Gerard, *and* Parkinson—Parkinson's *the last of importance (except* Salmon's*) which were exhibited in* England—*First copper-plates of plants.*

WOODEN CUTS.

AS we are now arrived at the period, when wooden cuts were about to be superseded by engravings on metal, PARKINSON's "Herbal" being the last of any importance in which they were used in *England*, it may not be incongruous to our plan to notice the origin and progress of that art, which contributed not a little to facilitate the knowledge of plants. Rude as these representations were, compared with the elegance of modern times, yet, in an age when specific distinctions were not fixed, and the diagnostic of the plant de-

pended

pended ſo much on habit, they ſpoke to the eye, and often diſcriminated the ſubject, when the laboured deſcription failed.

It has been before obſerved, that SEGUIER is of opinion the firſt Herbal with wooden cuts was the "*Puch der Natur*," "The Book of Nature," printed at *Augſburgh*, in 1478, if not three years earlier. Theſe are thought to have paſſed into the HERBARIUS, printed at *Mentz* in 1484; from which book was compiled the ORTUS SANITATIS, printed at the ſame place in 1485; with improvements in the work in general, and better figures, by CUBA. Of this work ſome notice has before been taken, as the foundation of the Engliſh "Grete "Herbal," firſt printed here in 1516.

The HORTUS SANITATIS was tranſlated into various languages, and in ſome new-modelled, without concealing its origin, according to the fancy of different editors and printers; and paſſed through innumerable editions on the Continent; having been the popular book on the ſubject, as the "*Grete Herbal*" was in England, for fifty or ſixty years.

It does not appear that CUBA was publicly known as the author of the HORTUS SANITATIS, until EGENOLF, a bookſeller of *Frankfort*, gave an improved edition, with an entirely new ſet of figures, under the care of EUCHARIUS RHODION or ROESLIN, a phyſician of the ſame city, in 1533. *Egenolf's* book paſſed through various editions, until a better work was compoſed by DORSTEN, under the title of "*Botanicon*," in 1540, at *Frankfort*; in which the ſame figures were employed. They were uſed alſo in the "*Encyclopædia Medica*" of J. DRYANDER, in 1542; and in the ſucceeding year, in an edition of DIOSCORIDES, by *Hermann Ryff*, printed by *Egenolf*. Finally, ADAM LONICER, the ſon-in-law of *Egenolf*, having totally reformed the work of CUBA, employed them in his *Herbal*, printed in 1546. In ſucceeding editions, he introduced new figures, took others from TRAGUS to the number in the whole of 880, and compoſed a work, which paſſed through a great number of editions, and was not ſuperſeded in the preſent century, as appears by an edition printed

printed ſo lately as in 1723, and even in 1737.

We are informed by TRAGUS, that *Egenolf* ſpared no expence in the encouragement of artiſts to procure theſe icons, rude and imperfect as they appear to us. He ſecured to himſelf, by this means, the monopoly of printing *Herbals*, for a ſucceſſion of years; and acquired both fame and riches.

At length, theſe were all ſuperſeded by thoſe of BRUNSFELSIUS to his Herbal, printed in 1532; which were drawn from nature, and appear to have been the firſt that were worthy of notice. Theſe were, however, greatly excelled by FUCHSIUS, in 1542; whoſe figures, although only outlines, are uncommonly beautiful, and not leſs juſt. They conſiſt of five hundred figures in folio, of the moſt common and uſeful plants; and were copied, in a ſmaller ſcale, by many ſucceeding authors. TRAGUS took moſt of them into his "Hiſtory "of German Plants," to which he added many new ones, to the amount in all of 567. Thoſe of TRAGUS are little more

† than

than outlines; and, allowing for the time, they ſufficiently well expreſs the habit of moſt of the ſubjects.

Egenolf having ſet the example, printers, after this time, themſelves bore the expence of cutting the blocks; by which means, certain printers monopoliſed the printing of *Herbals*; and a kind of commerce between them and authors took place, and mutual exchanges were made for the uſe of each other's books. Among theſe, no one poſſeſſed at length a greater collection than the famous PLANTIN, of *Antwerp*; who recommended himſelf ſo highly by the excellency of his types, and mode of executing his works. Hence he became the common printer to ſeveral of the celebrated botanic writers of the ſixteenth century. When CLUSIUS publiſhed his French tranſlation of DODOENS, with *Loe*, at *Antwerp*, he gave figures copied from FUCHSIUS; all which *Plantin* bought. He afterwards acquired the figures cut for CLUSIUS's own works, and thoſe of LOBEL. DODONÆUS, beſides ſome new blocks, had the uſe of all the above in the "*Pemptades*," in 1584, which work contains 1300 figures. TABERNÆ-

MONTANUS

MONTANUS obtained the uſe of this collection, namely, thoſe of FUCHSIUS, CLUSIUS, LOBEL, and DODONÆUS; to which he added thoſe of MATTHIOLUS; inſomuch that his Herbal, printed at *Frankfort* in 1588, comprehends more than two thouſand figures. DALECHAMP, in his "General Hiſtory of Plants," printed about the ſame time, augmented them to near two thouſand ſeven hundred.

The fate of GESNER's excellent figures I can but briefly mention; it forms a mortifying, but curious anecdote, in the literary hiſtory of the ſcience. Of the fifteen hundred figures left by GESNER, prepared for his "Hiſtory of Plants," at his death, in 1565, a large ſhare paſſed into the "*Epitome Matthioli,*" publiſhed by CAMERARIUS in 1586, which contained in the whole 1003 figures; and in the ſame year, as alſo into a ſecond edition in 1590, they embelliſhed an abridged tranſlation of MATTHIOLUS, printed under the name of the "German Herbal." In 1609, the ſame blocks were uſed by *Uffenbach* for the Herbal of CASTOR DURANTES, printed at *Frankfort*. This publication, however, comprehends

comprehends only 948 of theſe icons, nearly another hundred being introduced of very inferior merit. After this period, CAMERARIUS the younger being dead, theſe blocks were purchaſed by *Goerlin*, a bookſeller of *Ulm*; and next ſerved for the "*Parnaſſus Medicinalis illuſtratus*" of BECHER, printed at that city in 1663; the ſecond part of which work contains all thoſe of the "*Epitome*," except ſix figures. In 1678, they were taken into a German Herbal, made up from MATTHIOLUS, by *Bernard* VERZASCHA, printed at *Baſil*; and ſuch was the excellency of the materials and workmanſhip of theſe blocks, that they were exhibited a ſixth time in the "*Theatrum Botanicum*," or Kräuterbuch of ZWINGER, being an amended edition of VERZASCHA, printed alſo at *Baſil* in 1696, with the addition of more than one hundred new blocks, copied from C. BAUHINE and TABERNÆMONTANUS; and finally, into a new edition of the ſame work, ſo late as the year 1744.

Thus did the genius and labours of GESNER add dignity and ornament to the works of other men, and even of ſome whoſe enmity he had experienced during his life-time.

 Beſides

Besides the above mentioned, GESNER left five volumes, consisting entirely of figures, which, after various vicissitudes, became the property of TREW, of *Norimberg*. Sensible that whether we view the extent of GESNER's knowledge and learning, or his singular industry, such must be the veneration for his character, that any of his remains must claim the attention of the curious, the possessor gratified the public, by the pen of Dr. SCHMIEDEL, with an ample specimen, published in 1753.

Thus far for foreigners. The rude icons of the "Grete Herbal," it has been observed, were evidently copied from those in the HORTUS SANITATIS; for that they were not the same tables, appears from the diminished size. Of the figures in TURNER's History, which amount to upwards of 500; the greater part are those of FUCHSIUS's octavo set; and the remainder, nearly 100, were new. LYTE printed his translation of DODOENS with *Loe*, at *Antwerp*, for the conveniency of his figures, which are also borrowed from FUCHSIUS; to which LYTE added about thirty new ones.

GERARD, in 1597, and JOHNSON, his "*Emacu-*

" *Emaculator*" afterwards, in 1633 and 1636, procured all the blocks from *Frankfort*, with which the Herbal of TABERNÆMONTANUS had been illuſtrated. JOHNSON by this means accumulated upwards of 2700 cuts.

The blocks for PARKINSON's "*Theatrum*," and his "*Paradiſus*," were, I apprehend, cut in *England*; and thoſe for the firſt ſeem to be copies from GERARD, though much inferior in execution. The laſt of the kind uſed in *England*, were a new ſet cut for SALMON's "Herbal," in 1710; except, I believe, thoſe for a very indifferent performance, under the name of "An Herbal," publiſhed ſince that time, in quarto.

The earlieſt copper-plates of plants on the Continent, are ſaid to be thoſe of *Columna* in his "*Phytobaſanos*," in 1592. In *England*, except ſome ſingle figures, and the few plates in the firſt edition of PLOT's "Oxfordſhire" in 1677, thoſe of the "*Hiſtoria Oxonienſis*" are the firſt exhibition of any great work; and of theſe, the graſſes are, to this time, perhaps unparalleled in the neatneſs and accuracy of the execution.

CHAP. 13.

The botanical Garden founded at Oxford *by* Henry *Earl of* Danby—Jacob Bobart *the first Intendant—Two editions of the* Catalogus Oxoniensis—*Account of the authors, the* Bobarts, Stephens, *and* Browne.

Dr. How, *some account of—His* Phytologia *the first* English FLORA, *or separation of* English *from exotic botany—The author's assistants in this work,* Stonehouse, Bowles, *and others*—How, *the editor of* Lobel's *posthumous* Illustrationes.

HORTUS OXONIENSIS.

HITHERTO Botany, however successfully it might have been cultivated by individuals in *England*, had received no encouragement from any public institutions; but the time was now arrived, when it acquired additional vigour and improvement from the foundation of a physic-garden at *Oxford*. These elegant and necessary aids to science had considerably multiplied since the first foundations of the

the kind, before noticed, in *Italy* and elſewhere. Several univerſities in the more northern and weſtern parts of *Europe* had procured the eſtabliſhment of gardens: *Paris*, in 1570; *Leyden*, in 1577; *Leipſic*, in 1580; *Montpelier*, in 1598; *Jena*, in 1628; and *Oxford*, in the year 1632. This laſt was owing to the munificence of HENRY *Earl* of *Danby*, who gave for this purpoſe five acres of ground, built green-houſes and ſtoves, and an houſe for the accommodation of the gardener; endowed the eſtabliſhment, and placed in it, as the ſupervifor, *Jacob* BOBART, a German from *Brunſwick*, who lived, as WOOD tells us, in the garden-houſe, and died there on February 4, 1679. A liſt of the plants was publiſhed, under the title of "CATALOGUS PLANTARUM *Horti medici* OXONIENSIS *Latino-anglicus et Anglico-latinus: alphabetico ordine.*" Oxon. 1648. 12°. pp. 54 and 51. DILLENIUS informs us, that BOBART drew up this catalogue. In the preface we are told the garden contained 1600 ſpecies, by which muſt be underſtood both exotic and indigenous, including varieties of each. The plants are barely enumerated, without any ſynonyms, or references to any author. The number of Engliſh ſpecies recited, extends to 600, or nearly. The copiouſneſs of this catalogue ſets the zeal and diligence of BOBART in a favourable light. Under his care, and that of his ſon, the garden of *Oxford* continued to flouriſh for many years.

The CATALOGUS OXONIENSIS was republiſhed in the year 1658, in a much improved ſtate, by the joint aſſiſtance of Dr. STEPHENS, Mr. *William* BROWNE, and the two BOBARTS, father and ſon, under the following title, " CATALOGUS HORTI BOTANICI OXONIENSIS, *alphabetice digeſtus, duas præterpropter, plantarum chiliadas complectens, priore duplo auctior, idemque elimatior, nec non etymologiis, qua Græcis, qua Latinis, hinc inde petitis, enucleatior: in quo nomina Latina pariter et Græca vernaçulis; et in ejus ſequiore parte, vernacula Latinis præponuntur. Cui acceſſere plantæ minimum ſexaginta ſuis nominibus inſignitæ, quæ nullibi niſi in hoc opuſculo memorantur. Curâ et operâ ſociâ Philippi* STEPHANI, *M. D. et Gulielmi*

Gulielmi BROUNE, *A. M. adhibitis etiam in conſilium D.* BOBERTO *patre, hortulano academico ejuſque filio, utpote rei herbariæ callentiſſimis.*" Oxon. 1658. 8°. pp. 214.

Of Dr. *Philip* STEPHENS, whoſe name ſtands firſt among the authors of this catalogue, we find little mention elſewhere, as eminent in botanical ſcience. He was born at the *Devizes* in *Wiltſhire*, and was firſt of St. Alban's Hall, *Oxon*; afterwards made Fellow of New College by the viſitors, and became Principal of *Magdalen Hall*. He died at *London* after the Reſtoration.

MERRET, without any notice of Dr. STEPHENS, expreſsly calls Mr. BROWNE the author of this *Catalogue*; and *Wood* ſays, that he had the chief hand in it. *William* BROWNE was a native of *Oxford*, became Bachelor of Divinity, and Senior Fellow of Magdalen College. He died in March 1678, aged about 50, and was buried in the outer chapel of his college.

In this enlarged edition, the authors have, in every inſtance where it was poſſible, not only adopted the ſpecifical appellations given by GERARD and PARKINSON to each

 plant,

plant, but quoted the page of their works. This is the firſt book, as far as I know, on the ſubject, printed in *England*, in which the latter of theſe circumſtances takes place. It is remarkable, that ſo obvious an aſſiſtance, after having been introduced by *Caſpar* BAUHINE in his "*Phytopinax*," ſhould be wanting in the "*Pinax*" itſelf. Had GERARD and PARKINSON retained, throughout their works, the exact ſynonyms of the authors from whom they transferred their plants, and quoted the pages, they would unqueſtionably have rendered their writings much more uſeful to poſterity, and have preſerved them from diſuſe and oblivion, for a much longer period. The ſame may be obſerved of Mr. RAY, who has totally neglected this valuable improvement. So novel was the practice, that the authors of the HORTUS OXONIENSIS thought it neceſſary to apologiſe for it, and ſhield themſelves under the authority of the "*Hortus Eyſtettenſis.*"

There are many dubious and ill aſcertained plants in this Catalogue; and thoſe marked as new, are almoſt wholly varieties. Engliſh Botany ſeems to have received little

or

or no acceſſion by it; and I am not aware of one indigenous plant firſt mentioned in this liſt.

The ſecond part, or alphabetical liſt of Engliſh names, is intended only to lead to the Latin generical term in the firſt part.

HOW.

Until this period, no attempts had been made in *England* to ſeparate the indigenous from exotic botany. It is true, Dr. JOHNSON, as before mentioned, had publiſhed local catalogues of the plants of certain diſtricts; but no one had eſſayed a general liſt or deſcription of the Engliſh plants alone, in the way of what is now called a *Flora:* a term, which, as far as I can find, was firſt adopted by *Simon* PAULI, for a catalogue of the plants of *Denmark*, publiſhed in 1648. It is to Dr. HOW that we owe the firſt ſketch of a work of this kind; and, though he does not entitle his book *Flora*, he yet mentions that term in his preface.

William HOW was born in *London* in the year 1619, and educated at Merchant Taylors ſchool. He became a commoner of St. John's college, *Oxford*, at eighteen: he

took

took his bachelor's degree in 1641, and that of maſter of arts in 1645; and entered on the phyſical line. It does not appear that he ever took his doctor's degree, though he was commonly called Dr. How. With many other ſcholars of that time, he entered into the king's army, and for his loyalty was promoted to the rank of captain, in a troop of horſe. Upon the decline of the royal cauſe, he proſecuted his ſtudies in phyſic, and practiſed in that faculty. He lived firſt in St. Lawrence Lane, and afterwards in Milk Street. He died about the beginning of September 1656, and was buried by the grave of his mother, in St. Margaret's church, *Weſtminſter*; leaving behind him, as Mr. *Wood* ſays, "a choice library of books of his faculty, and the character of a noted herbaliſt."

Dr. How's principal publication, and for which he is here recorded, bears the following title:

"PHYTOLOGIA BRITANNICA, *natales exhibens indigenarum Stirpium ſponte emergentium.*" Lond. 1650. 12°. pp. 133.

The plants are arranged in the alphabetical order of the *Latin* names, with one or two

two ſynonyms, taken, as beſt pleaſed the author, from various writers on the continent, as well as from GERARD, PARKINSON, and LOBEL. The place of growth to each plant is noticed, and the particular ſpots where the rare ones grow, are ſpecified. The liſt contains 1220 plants, which (as few moſſes and fungi are enumerated) is a copious catalogue for that time, even admitting the varieties, which the preſent ſtate of botany would reject.

The author of this little volume was unqueſtionably a man of very conſiderable learning, and had a ſtrong paſſion for the knowledge of plants; but his ſituation in life does not ſeem to have allowed him the opportunity of travelling into the various parts of *England*, to gratify his taſte in *Engliſh* botany, with which he was not critically and extenſively acquainted. Mr. RAY, in the preface to his "*Catalogus Plantarum Angliæ*," has given a liſt of more than thirty ſpecies in the "*Phytologia*," which have no title to a place as indigenous plants of *England*. Some of theſe being inhabitants of Southern

Southern Europe; others evidently the accidental outcaſts of gardens; and ſome, as certainly, miſtaken for other plants, as appeared from the impoſſibility of finding them in the ſpots which How had pointed out.

The rare plants were almoſt wholly communicated by his friends, Mr. STONEHOUSE, Dr. BOWLES, Mr. HEATON, Mr. LOGGINS, Mr. GOODYER, and others. He drew ſome from a manuſcript of Dr. JOHNSON, the editor of GERARD. I wiſh it were in my power to commemorate theſe perſons in a more ample manner, who, at an early period, contributed to extend and illuſtrate Engliſh botany. Mr. STONEHOUSE, in particular, has deſerved highly of the lovers of this ſcience. He appears to have travelled much in *England*, from his recording the plants diſcovered by him in many counties. In *Yorkſhire* he was particularly converſant; and, I conjecture, he lived at a place called *Darfield*, near *Barnſley*, in that county.

Dr. BOWLES, and Mr. GOODYER, are, I believe, the ſame perſons mentioned under the

the article of JOHNSON. Of Mr. HEATON, I ſhall take further notice in the ſequel of theſe anecdotes.

It has been obſerved, that ſome of LOBEL's papers fell into the hands of PARKINSON, and ſome into Dr. HOW's poſſeſſion. Theſe were the fragment of LOBEL's great work, which HOW publiſhed in 1655, under the ſubſequent title:

"*Matthiæ de* LOBEL, *M.D. botanographi regii eximii,* STIRPIUM ILLUSTRATIONES, *plurimas elaborantes inauditas plantas ſubreptitiis Joh.* PARKINSONI *rapſodiis (ex codice M: S. inſalutato) ſparſim gravatæ, ejuſdem adjecta ſunt ad calcem Theatri Botanici* Ἁμαρτήματα. *Accurante Guil.* HOW, *Anglo.*" Lond. 1655. 4°. pp. 170.

This work has been noticed under the article of LOBEL. It is ſufficient to obſerve here, that the notes which the editor has affixed, would almoſt perſuade the reader that he had publiſhed the work with a view to take an invidious retroſpect of PARKINSON's "Theatre." In the preface to the "*Phytologia*," and in that of this work, both written in a flowery and bombaſt ſtile, as well

as

as throughout the notes, he ſpeaks of PARKINSON in very contemptuous language, and repreſents him as having made LOBEL's obſervations his own, without acknowledgment. Whatever may have been the caſe in particular inſtances; the attack, on the whole, was uncandid; ſince PARKINSON, in the very title of his "*Theatre*," profeſſes to have made uſe of, and inſerted, Dr. LOBEL's notes, together with thoſe of Dr. BONHAM and others. In fact, there is a petulance and an acrimony in the ſtile, both of the author and of the editor of this work, which, howſoever exampled in the laſt age, is, happily, much leſs frequently the language of literature in the preſent.

CHAP. 14.

Some account of the Tradeſcants, *father and ſon—The firſt who formed a muſeum of natural hiſtory in this country—Account of* Tradeſcant'*s publication—The muſeum bequeathed to* Aſhmole.

The aſtrological herbaliſts: Robert Turner, Culpepper, *and* Lovel—*The laſt the moſt reſpectable of the ſect in that time—Account of his* Pambotanologia —Pechey's Herbal — Salmon—*An account of his* Herbal.

TRADESCANT.

ALTHOUGH it does not appear that the TRADESCANTS contributed materially to amplify what is more eſpecially meant by Engliſh Botany, or the diſcovery and illuſtration of the plants ſpontaneouſly growing in *England:* yet, in a work devoted to the commemoration of Botaniſts, their name ſtands too high not to demand an honourable notice; ſince they contributed, at an early period, by their garden

garden and museum, to raise a curiosity thas was eminently useful to the progress and improvement of natural history in general.

John TRADESCANT was by birth a Dutch man, as we are informed by A. *Wood*. On what occasion, and at what period, he came into *England*, is not precisely ascertained. He is said to have been, for a considerable time, in the service of Lord Treasurer SALISBURY and Lord WOOTON. He travelled several years, and into various parts of *Europe*; as far eastward as into *Russia*. He was in a fleet that was sent against the *Algerines* in 1620, and mention is made of his collecting plants in *Barbary*, and in the isles of the Mediterranean. He is said to have brought the *trifolium stellatum Lin.* from the isle of *Fermentera*; and his name frequently occurs in the second edition of GERARD by JOHNSON; in PARKINSON'S "Theatre of Plants," and in his "Garden of Flowers," printed in 1656. But I conjecture that TRADESCANT was not resident in *England* in the time of GERARD himself, or known to him.

He appears however to have been established

bliſhed in *England*, and his garden founded at *Lambeth*; about the year 1629 he obtained the title of gardener to *Charles* I. TRADESCANT was a man of extraordinary curioſity, and the firſt in this country, who made any conſiderable collection of the ſubjects of natural hiſtory. He had a ſon of the ſame name, who took a voyage to *Virginia*, from whence he returned with many new plants. They were the means of introducing a variety of curious ſpecies into this kingdom; ſeveral of which bore their name. *Tradeſcant's Spiderwort*, *Tradeſcant's Aſter*, are well known to this day; and LINNÆUS has immortalized them among the Botaniſts, by making a new genus, under their name, of the *Spiderwort*, which had before been called *Ephemeron*. His Muſeum, called *Tradeſcant*'s Ark, attracted the curioſity of the age, and was much frequented by the great, by whoſe means it was alſo much enlarged, as appears by the liſt of his benefactors, printed at the end of "his MUSEUM TRADESCANTIANUM;" among whom, after the names of the king and queen, are found thoſe of many of the firſt nobility.

This ſmall volume, the author entitled

"MUSEUM TRADESCANTIANUM; or, a "Collection of Rarities preserved at South "*Lambeth*, near *London*. By *John* TRA-"DESCANT." 1656, 12°. It contains lists of his birds, quadrupeds, fish, shells, insects, minerals, fruits, artificial and miscellaneous curiosities, war instruments, habits, utensils, coins, and medals. These are followed by a catalogue, in English and Latin, of the plants of his garden, and a list of his benefactors. The reader may see a curious account of the remains of this garden, drawn up in the year 1749, by the late Sir *William* WATSON, and printed in the 46th volume of the Philosophical Transactions. Prefixed to this volume were the prints of both father and son; which, from the circumstance of being engraved by HOLLAR, has rendered the book well known to the collectors of prints, by whom most of the copies have been plundered of the impressions.

In what year the elder TRADESCANT died, is not certain, but his print abovementioned represents him as a man advanced in age.

The son inherited the *museum*, and bequeathed

queathed it by a deed of gift to Mr. ASHMOLE, who lodged in *Tradeſcant's* houſe. It afterwards became part of the *Aſhmolean muſeum,* and the name of TRADESCANT was unjuſtly ſunk in that of *Aſhmole.* John, the ſon, died in 1662. His widow erected a curious monument, in memory of the family, in *Lambeth* church yard, of which a large account, and engravings from a drawing of it in the *Pepyſian* library at *Cambridge,* are given by the late learned Dr. DUCARREL, in the 63d volume of the *Philoſophical Tranſactions**.

R. TURNER, CULPEPPER, and LOVELL.

The influence of Aſtrology in Phyſic and Botany, was far from being worn out in the middle of this age. By the credulity and ſuperſtition of ſome, and the diſhoneſty of others, it ſtill maintained its ground. Se-

* The name TRADESCANTIA was firſt applied by RUPPIUS, a German, in his *Flora Jenenſis,* to a plant introduced into the Engliſh gardens by TRADESCANT himſelf, and ſufficiently known by the appellation of *Tradeſcant's Spiderwort,* to which genus LINNÆUS has ſince reduced ſix other ſpecies.

 veral

veral phyſicians, and other men of learning, ſhewed ſome bias towards it. Many practitioners of an inferior claſs, and numerous empirics, were ſtill advocates for aſtrological influence in the preparation and application of ſimples.

There is an Herbal written by *Robert* TURNER, who calls himſelf *Botanologiæ Studioſus*, under the title of "BOTANO-"LOGIA, the Britiſh Phyſician; or, The "Nature and Vertues of Engliſh Plants; "exactly deſcribing ſuch as grow naturally "in the land, with their ſeveral names, "Greek, Latin, or Engliſh; natures, places "where they flouriſh, and are moſt proper "to be gathered; their degrees of tempera-"ture, applications, and vertues, phyſical "and aſtrological uſes treated of, &c." London, 1664, 12°. But, of the aſtrological herbaliſts, *Nicholas* CULPEPPER ſtands eminently forward. His "Herbal," firſt printed in 1652, whichc ontinued for more than a century, to be the manual of good ladies in the country, is well known; and, to do the author juſtice, his deſcriptions of common plants were drawn up with a clearneſs

clearneſs and diſtinction that would not have diſgraced a better pen.

Yet there is one author of this order, whoſe reſpectability might exempt him from total oblivion. *Robert* LOVELL's "compleat "Herbal," although ſaid to be written by him whilſt a young man, is of ſo ſingular a complexion, as to merit notice in a work of this kind, were it only to regret the miſapplication of talents, which demonſtrate an extenſive knowledge of books, a wonderful induſtry in the collection of his materials, and not leſs judgment in the arrangement. The firſt edition was printed in 1659; the ſecond in 1665, in 8°. at *Oxford*, pp. 672, exclusive of the introduction of 84 pages, and bears the following title, "PAMBOTANOLOGIA: *ſive Enchi-* "*ridion Botanicum*; or, A compleat Her- "bal; containing the ſum of antient and "modern authors, both Galenical and "Chymical, touching trees, ſhrubs, plants, "fruits, flowers, &c. in an alphabetical "order, wherein all that are not in the "phyſic garden in *Oxford* are noted with "aſteriſks. Shewing their place, time,

" names, kinds, temperature, virtues, uſe, " doſe, danger, and antidotes; together " with an introduction to herbariſme, &c. " an appendix of exotics, and an univerſal " index of plants, ſhewing what grow wild " in *England*; 2d edition with additions." *Oxford*, 1665, 12°.

To thoſe whoſe curioſity leans that way, it may not be eaſy to direct them to a more conciſe, or more perfectly methodical arrangement of ſimples, according to the Galenical principles of the four elements, temperaments, and qualities, than may be met with in the introduction to this book.

The arrangement of the matter in the work itſelf is according to the alphabet of the Engliſh names; to which is ſubjoined the place of growth, the time of flowering, then the name in Greek, and the Latin officinal term. There are no deſcriptions of the plants; but the qualities and uſes of each are collected from a profuſion of authors, and applied to all the ſpecies under each generical term; the form in which the medicine ſhould be given, the authority for each carefully cited, and the officinal compounds

compounds into which they enter aſſiduouſly noticed. The author includes ſimples, both of exotic and of indigenous growth.

He profeſſes to have cited near two hundred and fifty authors, of which he gives the liſt. At p. 482 begins an appendix on the drugs of the Eaſt and Weſt Indies, extracted from the Arabians, and from HERNANDEZ. A copious index of names to all the plants of his "Herbal," with the ſynonyms; eſpecially of the older authors; of ſuch as are mentioned in TRADESCANT; BAUHINE'S *Pinax*; of thoſe which are in the foreign botanical gardens, and not in that of *Oxford*; and laſtly, of thoſe in the PHYTOLOGIA BRITANNICA. The work concludes with a large index of diſeaſes, with the appropriate remedies from the ſimples of his work. In his catalogue of authors, he gives the number of figures contained in their works, which I tranſcribe as a matter of curioſity, that cannot fail to gratify the botanical reader *.

PECHEY.

* *Apollinaris ſ. Albertus,*	- -	141
Alpinus, Proſper	-	46
Bauhinu , J.	-	3547

PECHEY.

After the recital of CULPEPPER and LOVELL, I cannot refuſe admittance to an author of more reſpectability, though not deeply ſkilled in botanical knowledge.

" The compleat Herbal of Phyſical " Plants; containing all ſuch *Engliſh* and " foreign herbs, and ſhrubs, and trees, as " are uſed in phyſic and ſurgery. By *John*

Brunsfelſius - -	288
Camerarius, -	1003
Cluſius, Rariores -	1135
——— *Exotica*, - -	194
Columna, - -	205
Cordus, - -	272
Dodonæus, - -	1305
Durantes, - -	879
Eyſtettenſis Hortus -	1083
Fuſchſius, - -	516
Johnſon's Gerard, -	2730
Lobell, -	2116
Lonicerus, -	833
Matthiolus, - -	957
Parkinſon, -	2786
Rauwolf, - -	42
Renealme, - -	42
Ruellius, - -	350
Tragus, - -	567

" PECHEY,

" PECHEY, M. D. fellow of the college of " physicians." 8°. 1694; reprinted at *Amsterdam* the same year, and in 1707. The descriptions, which are short, are taken from RAY's history; the virtues from a variety of authors. The natural places of growth of the English plants are specified; but the author betrays his want of botanical knowledge, by enumerating several indigenous as exotic plants. PECHEY was the first who introduced into use the *casumunar*; of which he is said to have made a secret, and considered it as a corrector of the *Peruvian bark.*

In the same year was published, " PHILOBOTANOLOGIA: *s. Historia Vegetabilium sacra*; or, A Scriptural Herbal. By *William* WESTMACOTT." 8°. 1694. Not having seen this volume I can give no further account of it.

SALMON.

If my readers will excuse the anachronism, I am here tempted to anticipate the name of an author, the complexion of whose writings

writings renders it not improper to notice him after CULPEPPER and LOVELL; although in the time he lived, the influence of aſtrology had loſt ſtill more of its power. To the faſtidious critic in Botany, it might need ſome apology, that I introduce into theſe anecdotes the name of SALMON; well known as a multifarious writer, and author of numerous publications in phyſic, all of the empirical caſt. I confeſs, however, I could not paſs over, in total ſilence, a writer to whom, although no praiſe can be due as a botaniſt, yet the commendation of induſtry ought not to be withheld from a man who could beſtow twenty years labour, in the compilation of "an Herbal" of 1296 pages, in folio. I will recite the title, which will ſufficiently ſhew the nature of his work.

"The ENGLISH HERBAL; or, Hiſ-
"tory of Plants; containing, 1. Their
"names, *Greek*, *Latin*, and *Engliſh*. 2.
"Species, or various kinds. 3. Deſcrip-
"tions. 4. Places of growth. 5. Times
"of flowering and ſeeding. 6. Qualities
"or properties. 7. Their ſpecifications.
"8. Preparations, Galenic and Chymic.

 "9. Virtues

"9. Virtues and uſes. 10. A compleat "*florilegium* of all the choice flowers culti- "vated by our floriſts, interſperſed through "the work, in their proper places, where "you have their culture, choice, increaſe, "and way of management, as well for pro- "fit as delectation, adorned with exquiſite "icons, or figures of the moſt conſiderable "ſpecies. By *William* SALMON, M. D." *London*, fol. 2 vol. 1711.

The order of SALMON's book is alphabetical, and, as it is a work of mere compilation, he profeſſes to have conſulted all the botanical authors of repute, and enumerates the names of ſuch. His deſign was to treat on medicinal herbs principally. As a botanical work it is beneath all criticiſm; the errors in this way being enormous, both in multitude and degree. In detailing the powers of ſimples, he follows the Galenic terms of expreſſion uſed by the writers of the preceding century, and diſtributes, with a laviſh hand, extraordinary and numerous powers to almoſt every herb he deſcribes. Excluſive of his induſtry, ſome merit is due to SALMON for the regular arrangement of his

his ſubjects, ſubordinate to his method; qualities which, under the direction of more ſkill in Botany, and a ſounder judgement in diſcriminating the properties of ſimples, might have enabled him to have executed more effectually what ſeems to have been his purpoſe, that of ſuperſeding the Herbals of GERARD and PARKINSON, in which he totally failed. His tables, I have noticed heretofore, in ſpeaking on wooden cuts. But from theſe authors I return to writers of dignity and importance; and, with peculiar ſatisfaction, to the view, eſpecially, of a character, from whoſe penetrating genius, and perſevering induſtry, not Botany alone, but Zoology, may date a new æra. On this occaſion I ſingularly lament, that I am not furniſhed with any new materials to illuſtrate the life of RAY; of whom it may with truth be maintained, that in theſe branches of natural hiſtory, he became, without the patronage of an *Alexander*, the *Ariſtotle* of *England*, and the *Linnæus* of the time.

CHAP. 15.

Retrospective view of botanical science in the period immediately antecedent to Ray—*A detailed account of the life and writings of* Ray—*His* Catalogus Cantabrigiensis—Ray's *three first botanical tours—Appendixes to the* Cambridge Catalogue—*Foreign travels—Fourth tour in* England—*Elected fellow of the Royal Society.*

RAY.

IF we here take a retrospective view of the progress of botany during the first period of the seventeenth century, we find that, however particular individuals, both in *England* and on the continent, might have laboured in its advancement, it was not, on the whole, in a flourishing state, either here, or in any other part of *Europe*. From the time of the BAUHINES, even to that of RAY, its progress as a science was slow. The Remains of *John* BAUHINE, his "*Historia Plantarum Universalis*," printed in 1650, in three large folio

folio volumes, at the expence of 40,000 florins, defrayed by F. L. *à Graffenreid*, was the principal performance on the continent, and that indeed was invaluable. It is a monument of learning and induſtry, of which few examples can be expected in any one age. That which GESNER performed for zoology, *John* BAUHINE effected in botany. It is, in reality, a repoſitory of all that was valuable in the ancients, in his immediate predeceſſors, and in the diſcoveries of his own time, relating to the hiſtory of vegetables, and is executed with that accuracy and critical judgment which can only be exhibited by ſuperior talents.

The obſtacles to the improvement of botany were various. *Europe* had been involved in war, the perpetual enemy to free intercourſe among the learned; and to commerce, which is ever friendly to natural ſcience. Simples were neglected in phyſic, for medicines drawn from chymiſtry. Even alchymy yet employed the induſtry of many in every nation of *Europe*. Botanical gardens, although ſeveral, both public and private, had been eſtabliſhed, did not, however, flouriſh.

flouriſh. The *Indies* had not yet poured in their treaſures with that liberal hand which was ſoon after experienced. Even the paſſion of the floriſt for varieties aſſiſted in depreſſing the genuine ſpirit of the botaniſt. But the time was now approaching, when botany was about to receive a capital advantage and embelliſhment, by the introduction and eſtabliſhment of *ſyſtem*; of the riſe and progreſs of which, it will not be incongruous to my plan to give a ſhort account, ſince this great revolution formed a new æra in the hiſtory of the ſcience. As the revival of it, however, did not take place till the time of Mr. RAY and Dr. MORISON, I will poſtpone what I have to ſay on this ſubject, till I have given ſome account of the writings of thoſe juſtly celebrated naturaliſts, by whoſe labours ſyſtem itſelf was reſtored and improved.

The earlieſt anecdotes of Mr. RAY, to which I can refer, are ſome brief outlines of his life, in the "Compleat Hiſtory of Europe for the year 1705." A more connected account of this learned and excellent man may be ſeen in the "General Dictionary,"

nary," and the "Biographia Britannica;" but the moſt detailed relation is that of Dr. SCOTT, publiſhed in 1760, from materials collected by Dr. DERHAM. This is well abridged in the *Biographical Dictionary*. It is much to be regretted, that our curioſity has not been more amply gratified than by theſe ſhort and imperfect memoirs.

A more circumſtantial narrative of the life of Mr. RAY would, even at this diſtance of time, be a valuable acceſſion to biography, and highly grateful to thoſe, who are ſenſible of the great improvements which he gave to the ſcience of natural hiſtory in general; nor could ſufficient juſtice be done to his manifold talents, diſcoveries, and writings, but by a pen of the firſt eminence in biographical literature.

The limits of my plan will not allow of more than a general detail of the principal events of his life, as connected in chronological order with his writings.

John *Wray*, or, as he always ſpelt his name after the year 1669, RAY, was born at *Black Notley*, near *Braintree*, in *Eſſex*, Nov. 29, 1628

1628. His father, though in so humble a situation as that of a blacksmith, sent his son to the grammar-school at *Braintree*; and in 1644, entered him at Catherine Hall, in *Cambridge*; from whence he removed, in less than two years, to Trinity College, where the politer sciences were more cultivated. Dr. BARROW was his fellow pupil, and intimate friend, and, on account of their early proficiencies, both were the favourites of their learned tutor, Dr. DUPORT. He was chosen minor-fellow of Trinity, in 1649; in 1651, was made Greek lecturer of the college; in 1653, mathematical lecturer; and in 1655, humanity reader. These appointments were sufficient testimonies of his talents and abilities at this early period. He afterwards passed through the offices of the college, and became tutor to many gentlemen of honourable birth and attainments, who gave him due praise and acknowledgments for his watchful care of them. He also distinguished himself, while in college, as a sensible and rational preacher, and a sound divine. As his favourite study was the works of God, he laid, at this time, in his college lectures, the foundation of his "Wisdom of God in the Creation," and of his "Three Physico-theological Discourses;" which were afterwards so well received by the public.

At the period when Mr. RAY turned his attention to the study of nature, the knowledge of plants was not highly superior to the state in which TURNER had found it, in the same place, more than a century before. In this study RAY could find no master. I am not able to say, that a single publication, of a scientific nature, on the subject of plants, had ever appeared at *Cambridge*; for *Maplet*'s "Green Forest" will scarcely be thought worthy of that appellation. *Oxford* had, indeed, not only experienced the benefit of private encouragement, but of public munificence, in the establishment of a Garden. But at the sister university Mr. RAY stood alone, himself indeed an host! Self-taught as he was, and full of ardour, he so forcibly displayed the utility of botanical knowledge, and its intimate connection with the arts, and conveniences

niences of life, independent even of thoſe charms, which the views of nature ever afford to contemplative minds, that he ſoon made it an object of attention; and numbered among his aſſociates in theſe ſtudies, Mr. NID, a ſenior fellow of his own college, Mr. *Francis* WILLUGHBY, and Mr. *Peter* COURTHOPE. The firſt of theſe gentlemen became his inſeparable companion; but he had the misfortune to deplore his death, a little time before the publication of his firſt work, which came out under the title of "CATALOGUS PLANTARUM CIRCA CANTABRIGIAM NASCENTIUM." *Cantab.* 1660." pp. 182. *cum Indicibus, &c.* pp. 103. 12°.

This little volume contains all the plants which the author had obſerved ſpontaneouſly growing in the neighbourhood of *Cambridge*, amounting to 626, all varieties and dubious plants excluded. The number is ſmall, when compared with many modern catalogues; but not ſo, when it is recollected, that, at that period, a very few of the *Cryptogamia* claſs, and not many of the *Graminaceous* tribe, had been investigated.

The plants are diſpoſed in the alphabetical order of the *Latin* names; and the ſynonyms of the four principal authors then in uſe given at length. Theſe are GERARD and PARKINSON, and the two BAUHINES; nor are others wanting, when characteriſtic of the plant. Prefixed is a liſt of the authors, ſo accurately and inſtructively drawn up, as not to have loſt its utility to this day. Mr. RAY has interſperſed many ſelect obſervations, on the medicinal and œconomical uſes of the plants; on the ſtructure of the flower; on varieties: and has not only deſcribed ſome new plants, diſcovered by himſelf, but given accurate diſtinctions of many, before imperfectly known. Subjoined, the reader finds an index of the *Engliſh* names, preceding the *Latin*; an index, ſpecifying the particular places of the more rare plants; then, a copious etymology of the names, and an explanation of the terms uſed in the ſcience. In fine, he has done every thing to facilitate the labour

of

of the ſtudent in this part, as in the former to inſtruct and entertain the more erudite reader.

I have been the more diffuſe on this ſmall volume, as the author has obſerved nearly the ſame plan, in his ſubſequent catalogues, and SYNOPSIS. *Moles parva, Vis magna.* When the time in which this publication was made, and the meagre ſtructure of preceding catalogues is conſidered, I may ſafely appeal to modern judges, whether this was not an extraordinary production. Few local catalogues had been publiſhed at home; and, I believe, not one abroad, that diſplayed any thing like a comparable ſhare of ſcience and erudition, ſo aptly united.

Among the variety of notes in this catalogue, there is one, poſſibly not of public notoriety. Mr. RAY informs us, that the people of *Norwich* had long excelled in the culture and production of fine flowers; and that in thoſe days, the floriſts held their annual feaſts, and crowned the beſt flower with a premium, as at preſent.

There can be no doubt that this volume met with the moſt favourable reception from the learned in this way; that it promoted the ſtudy of plants; and, by raiſing the reputation of its author, encouraged him to proſecute his ſtudies with vigour.

Theſe occupations, however, did not divert Mr. RAY from his object of entering into the miniſtry. He was, in Dec. 1660, ordained both deacon and prieſt, by Dr. *Sanderſon*, biſhop of *Lincoln*, and continued fellow of *Trinity College* till the Bartholomew act; which, as he did not ſubſcribe, neceſſarily ſuperſeded him. This event took place Sept. 18, 1662.

The deſire Mr. RAY had to extend his knowledge of *Engliſh* botany, had induced him, in the autumn of 1658, to take a journey, which he performed alone, through the midland counties of *England*, and the northern part of *Wales*, in ſearch of plants. This tour held him from Auguſt 9, to September 18. Of this, and of two other tours, Mr. RAY preſerved ſome ſhort memorandums, in which he has noticed his daily progreſs, ſome remarkable facts that occurred, ſome obſervations on the antiquities that he met with, and ſome of the rare plants.

plants. Dr. SCOTT has publiſhed theſe *Itineraries*, with his life.

In his ſubſequent journies, he was commonly accompanied by ſome friends of a congenial taſte; thus, in his ſecond tour, in the autumn of 1661, Mr. WILLUGHBY, and ſome other gentlemen, travelled with Mr. RAY into *Scotland*, through the counties of *Durham* and *Northumberland*, to *Edinburgh*, *Glaſgow*, and back through *Cumberland* and *Weſtmorland*. This journey held ſix weeks, from July 26, to Auguſt 30. In 1662, Mr. RAY, accompanied by Mr. WILLUGHBY, took his third and moſt extenſive *Engliſh* tour; through the middle counties of *England*, into *Cheſhire*; thence into *North Wales*, and through the middle *Welch* counties, into *Pembrokeſhire*, coaſting the ſouthern part, to *Bath* and *Briſtol*; thence to the *Land's End*, through *Somerſet* and *Devon*; returning through *Dorſetſhire*, *Wiltſhire*, and *Hampſhire*. They were abſent in this excurſion, from May 8, to July 18; and Mr. RAY gathered a plentiful harveſt, which afterwards enabled him to enrich his general "Catalogue of *Engliſh* Plants,"

 then

then in meditation; nor did he omit to avail himſelf of every opportunity, particularly at *Tenby*, in *Wales*, and in *Cornwall*, of deſcribing ſuch birds and fiſhes as were leſs frequent in other parts, preparatory to his intended publications in the zoological way.

In 1663 he publiſhed an *Appendix* to the "Cambridge Catalogue," containing emendations, and the addition of forty-two plants. And in 1685, came out another *Appendix*, with the addition of ſixty more, not noticed before; which were principally communicated by Mr. DENT, of *Cambridge*. Theſe little tracts are become very ſcarce. Thoſe who are curious to ſee what theſe additional plants were, may find them diſtinguiſhed from the others in Profeſſor MARTYN'S "*Plantæ Cantabrigienſes*."

Being now at liberty from the conſtraints and buſineſs of a college life, he was led to accompany Mr. WILLUGHBY, Mr. SKIPPON, and Mr. *Nathaniel* BACON, two of them his pupils, to the continent. Mr. RAY was abſent from April 18, 1663, to March 1665-6; during which time, they viſited

visited *France*, *Holland*, *Germany*, *Switzerland*, *Italy*; and extended their journey to *Sicily* and to *Malta*. The fruit of this expedition will afterwards appear.

On his return from the continent, he spent the summer of 1666 between his friends in *Essex* and *Sussex*, and in reading the publications which had appeared in *England* during the three years of his absence. The winter passed in reviewing and arranging the museum of his friend and pupil, Mr. WILLUGHBY, rich in animal and fossil productions; in arranging his own catalogues for his general list of *English* vegetables; and in framing the tables for Dr. WILKINS's "Real, or Universal Character."

In the summer of 1667, Mr. RAY, accompanied by his much-honoured friend, Mr. WILLUGHBY, made his fourth excursion into the distant counties. They left *Middleton Park* on June 25, and took their route to the *Land's End*, through the counties of *Worcester*, *Gloucester*, and *Somerset*; and returned through *Hants* to *London* on September 13. In this journey, besides the pointed

pointed objects of their pursuit, they took notes on the mines, and smelting, and on the method of making salt; and Mr. RAY did not omit to make, as he had done before, ample additions for his collections of proverbs and of local *English* words.

On Nov. 7, of this year, he was chosen fellow of the Royal Society, and was prevailed on by Bishop WILKINS to translate his "Real Character" into *Latin*. This he performed, though it was never published; and the manuscript is extant in the library of the Royal Society. The latter end of the year, and the beginning of 1668, he spent with gentlemen who had all been his pupils at Trinity; Mr. BURREL, and Mr. COURTHOPE, at *Danny*, in *Sussex*; Sir *Robert* BARNHAM, at *Boston*, in *Kent*; and with Mr. WILLUGHBY, in *Warwickshire*. In the autumn of this year, he took his fifth journey, alone, into *Yorkshire* and *Westmorland*, returning in September to *Middleton Hall*; and spent the winter with Mr. WILLUGHBY, then lately married.

CHAP. 16.

Account of Ray *continued—Makes experiments on the motion of the ſap*—Catalogus Plantarum Angliæ—*Sixth tour in* England—*Deceaſe of his friend* Mr. Willughby—*and of* Bp. Wilkins—Nomenclator Claſſicus — *His marriage* — His Obſervations topographical and moral, *&c. made in his foreign travels : with the* Catalogus Stirpium Exoticarum, *annexed.*

RAY.

ABOUT this time Dr. TONGE, Dr. BEAL, and ſome other philoſophical gentlemen, in *England*, were buſied in experiments relating to the motion of the ſap in trees. Among theſe alſo, in the ſpring of 1669, Mr. RAY and Mr. WILLUGHBY entered upon a ſet of the like experiments, and induced Mr. (afterwards Dr.) LISTER, to proſecute the ſame. Theſe experiments were made on the birch, the ſycamore or greater maple, the alder, the aſh, the haſel, cheſnut, walnut, and willow; of which the two firſt were

were found to be the beſt adapted to the purpoſe, from their bleeding moſt freely.

The experiments of Mr. RAY and Mr. WILLUGHBY, which were printed in the fourth volume of the *Philoſophical Tranſactions*, proved the aſcent and deſcent, as well as the lateral courſe, of the ſap; but theſe gentlemen declined giving any deciſive opinion, as to a real circulation upwards by the veſſels of the wood, and downwards by thoſe between the wood and the bark; which was the doctrine maintained ſoon after this time by GREW and MALPHIGI, and indeed afterwards adopted by Mr. RAY himſelf.

This doctrine of the circulation of the ſap, I need ſcarcely remark, gave way to the experiments of Dr. HALES and others; which teaches, that the ſap riſes and falls, in the ſame ſyſtem of veſſels, as it is affected by the joint operations of air and warmth. Yet there have not been wanting ingenious men of late years, alſo, who, conceiving the analogy between animals and vegetables to be greater than is uſually imagined, and even that plants not only live, but feel, have

advanced

advanced it as ſtill probable, that there is a real circulation of the juices; the *ſuccus communis* riſing from the roots, and the *ſuccus proprius* deſcending towards them. Whether theſe phyſiologies will yield to the *prolepſis plantarum* of the LINNÆAN ſchool, time muſt evince.

When Mr. RAY was at *Cheſter*, in 1669, he availed himſelf of an opportunity of viewing a young porpeſs, and of attending the diſſection of it. Of the anatomical ſtructure of this animal, he communicated a circumſtantial account to the R. S. in 1671; and it was printed in the *Philoſophical Tranſactions*, N° 74 and 76.

In 1671, Mr. RAY wrote a paper, printed in the *Philoſophical Tranſactions*, N° 74, on the ſubject of "Spontaneous Generation," a point of philoſophy which had been much diſcuſſed, and to which ſome among the learned were yet attached. It appears from this paper, that he very early rejected this doctrine, and was confirmed in his opinion by the experiments of REDI.

We are now to reap the fruit of Mr. RAY's repeated journies into the various parts

parts of *England*, taken with a professed view, to ascertain the *loci natales* of all the native plants, more accurately than had yet been done; to investigate the more rare, and perchance to discover new ones. In each of these departments he had proved successful, and in this year drew up his "Catalogue," and dedicated it to his friend and Mæcenas Mr. WILLUGHBY, under the following title, "CATALOGUS PLANTARUM ANGLIÆ *et insularum adjacentium tum indigenas tum in agris passim cultas complectens.*" *Lond.* 1670, pp. 358. 8°.

This work is modelled after the *Cambridge* Catalogue in general, as to the order of the subject, except that the author has been much more sparing of the synonyms, from all authors but the four classial writers, GERARD, PARKINSON, and the two BAUHINES. Several new plants are described in this volume, and many doubtful ones discriminated, with that critical accuracy which so singularly marked his pen; and which had not before been seen in any *English* writer.

Hitherto the *cryptogamous* and *graminaceous*

ceous tribe, had engaged but little attention; and in this volume, theſe claſſes do not far exceed the number regiſtered in the *Cambridge* Catalogue. The whole number of plants in this liſt, amount to about 1050 only. This ſmall number had been owing to the extreme caution of Mr. RAY, not to admit any varieties to hold the place of ſpecies; and to exclude all others on doubtful authority. How, in his "*Phytologia*," has upwards of 1200; and MERRETT, in his "*Pinax*," upwards of 1400; certain proofs that the authors had not ſufficiently ſtudied the nicer diſtinctions, which guided the judgment of Mr. RAY; and as a proof, it may be obſerved, that many of their plants are to this day undiſcovered. Senſible as Mr. RAY was of the errors of MERRETT's "*Pinax*," he purpoſely omitted quoting it, as he writes to Dr. LISTER, that he might avoid that cenſure of it, which could not properly have been withheld, had he given his impartial opinion of that performance.

In this year, he informs Dr. LISTER, that he had, what he thought, a moſt liberal offer, of one hundred pounds a year, and

all

all his expences defrayed, to accompany three young gentlemen abroad. But he declined it, although he much wiſhed to have taken a review of the alpine plants. Indiſpoſition had ſome ſhare in this refuſal, and we find that in the next ſpring, 1671, he ſuffered much from a jaundice. He was ſo far recovered, however, before *July*, as to be able to ſet off on his ſixth journey, in which he took with him *Thomas* WILLISEL, an unlettered man, but one, whoſe love for plants, and his zeal and aſſiduity in collecting them, merits commemoration. They travelled through *Derbyſhire*, *Yorkſhire*, and all the northern counties, as far as to *Berwick*, and back through the biſhoprick of *Durham*.

In the ſame year died, to the unſpeakable loſs and grief of Mr. RAY, his moſt valuable friend *Francis* WILLUGHBY, Eſq; on *July* 3d, in the 37th year of his age. The ſtricteſt intimacy had ſubſiſted between them, from the time of their being fellow collegians; and it was cemented by a congeniality of taſte, which not unfrequently forms a ſtronger bond of union, than the ties

ties of blood. Mr. WILLUGHBY had imbibed, very early, a ſtrong taſte for the ſtudy of the animal kingdom, and had made extraordinary collections for compleating the "Hiſtory of Birds and Fiſhes;" in which he had ever been aſſiſted by his friend Mr. RAY; who experienced his high attachment and confidence, in being left one of his executors, and charged with the education of his two ſons, the eldeſt of whom, was not four years of age. To this care he liberally annexed an annuity of ſixty pounds per annum for life, which was ever regularly paid.

Immediately after this melancholy event, he deſiſted from journeying again into the weſtern counties, as he had intended; and refuſed an invitation from Dr. LISTER, to live with him at *York*; in order to give himſelf up to the faithful diſcharge of his truſt.

For the uſe of theſe young gentlemen, Mr. RAY drew up, in 1672, his *Nomenclator Claſſicus*, induced thereto by obſerving the multitude of errors in the names of plants and animals, in the manuals of daily uſe. This compilation had authority

 enough

enough to recommend itſelf to ſubſequent writers of dictionaries and lexicons, and has been reprinted ſeveral times.

On *November* 19th, 1672, he ſuſtained, in the death of Biſhop WILKINS, the loſs of another of his beſt friends. For this candid, ingenious, and learned man, he had a ſincere eſteem and veneration.

In the lot of human life, ſuch chaſms are not eaſily filled up after the age of forty-five. It is however not unreaſonable to conjecture, that theſe privations added ſtrength to his motives for domeſtic retirement, and accelerated at leaſt, that connexion he made the next year, when he married Margaret the daughter of Mr. *John Oakely*, of *Launton*, in *Oxfordſhire*. They were married in the church of *Middleton*, on *June* the 5th, 1673.

In the ſame year Mr. RAY gave to the public the fruit of his foreign travels, under the title of, "Obſervations, topographical, "moral, and phyſiological, made in a jour-"ney through part of the Low Countries, "*Germany*, *Italy*, and *France*." *London*, 1673. 8°. pp. 499.

The great object of accompanying his three

three aſſociates in this tour, was, the enlargement of his knowledge in natural hiſtory, and particularly in the vegetable kingdom; and the great number of plants obſerved and collected by him, exceeded, as he informs us, his expectation: not that any opportunities eſcaped him of deſcribing the birds and fiſhes of the ſeveral countries they paſſed through, in aid of Mr. WILLUGHBY's plans. His notes concerning thoſe of *Germany*, were unfortunately loſt. The volume before us, however, is by no means confined to natural hiſtory. Mr. RAY treats on the manners of the people, and expatiates often on the excellencies and defects of the ſeveral governments, particularly of the cities on the continent, and on the ſtate of the academies and univerſities. He does not omit to notice the antiquities that occurred and of thoſe at *Rome*, he gives a very methodical account. Beſides many miſcellaneous remarks on various other parts of natural hiſtory, he has taken occaſion to make a digreſſion, which, at that time, muſt have been of a very intereſting nature, on the moſt remarkable places,

 where

where petrified ſhells and figured foſſils are found, both in *England* and elſewhere; and on the various opinions of authors, relating to the origin of theſe bodies. He freely declares his ſentiments, that they are the remains of once-organized bodies, in oppoſition to thoſe who imagined them to be the product of what they called a *plaſtic* power. He afterwards confirms his poſitions, by additional arguments recited in a letter to Dr. ROBINSON. See *Letters*, p. 165.

In the courſe of their journey, he every where notices thoſe plants that are not natives of *England*, and gives copious catalogues of them. They ſpent in the whole, ſix months at *Geneva*, which gave Mr. RAY an opportunity of informing himſelf largely, relating to the plants of *Switzerland*, particularly thoſe of Mount *Saleve*, the *Dole*, and of Mount *Jura*. He even diſcovered ſome that were unknown to the preceding botaniſts, although theſe were the regions of GESNER, and the BAUHINES.

The celebrated HALLER, even ranks him

him among thoſe who made large acceſſions to the Botany of that country, and gives the ſtrongeſt teſtimony of his ſkill, fidelity, and judgment, in diſcriminating, deſcribing, and extricating the plants of that fruitful region.

To the end of theſe "Obſervations," is affixed an alphabetical liſt of the plants mentioned in the body of the book, under the title of "CATALOGUS STIRPIUM IN EXTERIS REGIONIBUS, *à nobis obſervatarum, quæ vel omnino vel parcè admodum in Anglia ſponte proveniunt.*" pp. 115.

In the arrangement he cites the ſame authors for ſynonyms as in his preceding catalogues, and occaſionally introduces obſervations on the qualities and uſes.

CHAP. 17.

Account of Mr. Ray *continued—His various erudition—Collection of* Engliſh Proverbs—*Collection of* Engliſh Words—*Second edition of the* Catalogus Plantarum Angliæ—*Publiſhes* Willughby's Ornithology, *both in* Latin, *and in* Engliſh—*Engaged by the* R. S. *to make experiments in natural hiſtory—Removal to* Black Notley, *in* Eſſex—*Publiſhes* Willughby's Icthyology.

RAY.

THE talents of Mr. RAY were not confined to natural hiſtory. He had a reliſh, among other departments of literature, for philological enquiries, and the genius of the *Engliſh* language had engaged much of his attention. Of his purſuits in this way, he has left memorials, which have extended his reputation beyond the ſphere of natural hiſtory, and made him known to the learned world in general.

I refer to his " Collection of *Engliſh* " Proverbs," and to his " Collection of

"*English* Words." The foundation of these publications was laid in his various tours through the different parts of *England*. His " Proverbs" were finished for the press in 1669, but not published till 1672, and a second edition, much enlarged, in 1678, under the following title: " A COLLEC-" TION OF ENGLISH PROVERBS, digested " into a convenient method for the speedy " finding one upon occasion; with short " annotations. Whereunto are added local " proverbs, with their explications, old pro-" verbial rhythmes, less known, or exotic " proverbial sentences and *Scottish* pro-" verbs. Enlarged by the addition of many " hundred *English*, and an appendix of *He-*" *brew* proverbs, with annotations and pa-" rallels." *Cambridge*. 8°. pp. 414.

It has been reprinted many times, and, I think, so lately as in the year 1768.

To collect these sententious maxims of knowledge, both of a moral, prudential, and even a jocular nature, has not been deemed unworthy employment, by men of eminent learning and intelligence. The Adagies of ERASMUS furnish a sufficient example of the eſtimation he gave them. They were an oral and traditionary kind of didactics, which bore a greater value before the diffuſion of knowledge by the uſe of printing; and, in oriental countries, are ſtill a favourite and uſual mode of inſtruction.

Of ſuch as have been handed down in *Britain*, from father to ſon, through numerous generations, Mr. RAY's collection contains an ample ſtore. It is, I believe, the principal in its way; and the author has interſperſed many notes, which illuſtrate the origin and ſenſe of theſe aphoriſtic leſſons, and throw no ſmall light on the manners and cuſtoms of various people.

In 1674, was publiſhed, his "COLLEC-"TION OF ENGLISH WORDS not gene-"rally uſed, with their ſignifications, and "original, in two alphabetical catalogues, "one of the northern, and the other of the "ſouthern counties. To which is added, "an account of the preparing and refining "ſuch metals and minerals as are gotten in "*England*." *London*, 12°.

This little volume is dedicated to his friend Mr. COURTHOPE, at whoſe ſuggeſtion,

tion, he tells us, it was undertaken, and who contributed largely to augment it. In the firſt edition was a catalogue of the *Engliſh* birds and fiſhes; but this was omitted in a ſubſequent improved and enlarged edition, in 1691, Mr. RAY having then projected his "*Synopſis Animalium.*"

This is one of thoſe philological collections, which tends to amuſe and gratify general curioſity, is of uſe, not only to ſtrangers and thoſe who travel, but to thoſe who ſtay much at home; while it contributes to enlarge the extent, and illuſtrate the conſtruction of the *Engliſh* tongue. Mr. *Thoreſby*, of *Leeds*, ſent to Mr. *Ray*, a large addition to this liſt in the year 1703, which was printed in his "Philoſophical Letters," by Dr. *Derham*.

In 1675, he communicated to the *Royal Society* ſome experiments, made, I believe, by Mr. WILLUGHBY, accompanied with his own obſervations, tending to aſcertain the true uſe of the *air-bladder* in fiſhes. They are ſuch as the preſent phyſiology of fiſhes have confirmed; and were printed in the Philoſophical Tranſactions, N° 115.

In

In the year 1677, his "Catalogue of "*English* plants" being out of print, he gave another edition, augmented with new obſervations, and the addition of 30 ſpecies of the more perfect plants, and 16 *funguſes*; ſeveral of theſe were new diſcoveries. He here alſo gives the figures of the *pentaphylloides fruticoſa, (potentilla fruticoſa Lin.)* and the *fungus phalloides (phallus impudicus Lin.)*

Mr. RAY continued, after his marriage, to reſide at *Middleton Hall,* where his engagements at this period of his life, were ſuch as called forth all the talents of his literary abilities, and demanded all his care as a faithful guardian. He was employed in a double duty, that of his truſt to the ſons of his late eſtimable friend, and of editor to the remains of their father, "On the Hiſ-"tory of Birds and Fiſhes." The *Ornithology* was firſt publiſhed, to which, as it conſiſted of looſe papers, written in *Latin,* and in an undigeſted ſtate, Mr. RAY gave method, and ſupplied, from his own obſervations, a large ſhare of valuable materials. It was publiſhed under the following title: "ORNITHOLOGIÆ LIBRI TRES: *in quibus, Aves*

Aves omnes hactenus cognitæ, in methodum naturis ſuis convenientem redactæ accuratè deſcribuntur. Iconibus elegantiſſimis et vivarum avium ſimillimis æri inciſis illuſtrantur. Totum opus recognovit, digeſſit, ſupplevit Johannes RAIUS." *Lond.* 1676, fol. pp. 307, t. 77, f. 353.

Mr. RAY tranſlated this work into *Engliſh*, and publiſhed it, with large additions, in 1678, with figures engraved at the expence of Mrs. WILLUGHBY. The execution of the figures was wholly inadequate to the merit of the work. Theſe occupations, however, did not prevent him from renewing a correſpondence with Mr. OLDENBURGH, ſecretary of the Royal Society, a learned German, who, after having reſided ſome time at *Oxford*, had been choſen into that office at the firſt eſtabliſhment of the ſociety. Mr. RAY, in the year 1674, was induced to engage, at the requeſt of the ſociety, with other diſtant members, to furniſh obſervations on the ſubjects of natural hiſtory, to be read at their meetings, the ſociety notwithſtanding the extreme diligence of the ſecretary, and ſome few others,

others, being, at this juncture, rather in a languiſhing ſtate.

On this occaſion he wrote ſeveral papers, of which ſome were afterwards printed in the *Philoſophical Tranſactions*. Among thoſe, which were not publiſhed, as we find by his letters, were the following. "On the Acid of Ants: On a Foſſil of the figured Kind, found in *Malta*, and known by the name of St. *Paul's Baſtoons Letters*, p. 120: On the *Trochites*: On Muſhrooms: On the Darting of Spiders: On the Seeds of Plants; and on the ſpecific Differences of Plants."

On the death of the mother of his friend, the Dowager Lady WILLUGHBY, and the removal of his ſons from under Mr. RAY's tuition, he retired, ſome time in the year 1676, to *Sutton Coſield*, about four miles diſtant from *Middleton Hall*, where he remained till Michaelmas 1677. He then made a ſecond removal to *Falkborne Hall*, near *Black Notley*; at which laſt place he built a houſe, and finally ſettled *June* 24, 1679.

Mr. WILLUGHBY's *Icthyology* remaining yet

yet unpublished, Mr. RAY, in 1684, arranged the materials, which had been left in a very imperfect and indigested state. Perhaps no one but Mr. RAY could have fulfilled this posthumous office; certainly no man so effectually, since Mr. RAY had not only himself entirely furnished Mr. WILLUGHBY with many, but even the remainder had chiefly been collected during their almost daily intercourse, and whilst travelling together.

He wrote the two first books himself; revised, methodised, and enlarged the whole; and sent it to the Royal Society; the members of which contributed to furnish the plates; and, by the assistance of Bishop FELL, it was printed at *Oxford*; the Royal Society being at the whole expence. It came out under the following title:

"*Francisci* WILLOUGHBEII, *Armig. De* HISTORIA PISCIUM, LIBRI *quatuor, jussu et sumptu* S. RAY. *Lond. editi. Totum opus recognovit, coaptavit, supplevit, librum etiam primum et secundum integros adjecit* J. RAIUS." Oxon. 1686. fol. pp. 343.

CHAP.

CHAP. 18.

Account of Mr. Ray *continued—Meditates the writing of his* General Hiſtory of Plants—Methodus Plantarum, *as introductory to that work—Two firſt volumes of the* Hiſtory, *in which are deſcribed near ſeven thouſand plants* —Faſciculus Stirpium—*Firſt edition of the* Synopſis Stirpium Britannicarum.

RAY.

MR. RAY being ſettled at *Black Notley*, and delivered from that anxiety which had attended him ſince Mr. WILLUGHBY's death, reſumed with great vigour his wonted ſtudy of plants; and, having already acquired a reputation that juſtified any expectation his friends might have formed, he, in compliance with their wiſhes, attached himſelf ſeriouſly to write "A General Hiſtory of Plants."

Preparatory to this great work, which he intended to arrange ſyſtematically, he put forth, in 1682, his "METHODUS PLANTARUM," enlarged, and improved, from the ſynoptical

ſynoptical tables, which he had printed in Biſhop WILKINS's "Real Character," in 1668. It bears the following title:

"METHODUS PLANTARUM NOVA *brevitatis et perſpicuitatis cauſâ ſynoptice in tabulis exhibita: cum notis Generum tum ſummorum tum ſubalternorum characteriſticis. Obſervationibus nonnullis de ſeminibus Plantarum et indice copioſo.*" Lond. 1682. 8°. pp. 166.

LINNÆUS, on what authority I know not, mentions an edition of this work, with the date of 1665, totally ſeparate from that of 1682; but as that is earlier than Biſhop WILKINS's Table, it is probably a miſtake.

The firſt principle of Mr. RAY, in this work, is to preſerve all plants together, as far as poſſible, in the natural characters, ariſing from conformity in the fructification, and in the general habit. Hence aroſe, with him, in common as with others, too great a neglect of the flower, and too much attention to the leaves. He adheres to the ancient diviſion of the vegetable kingdom, into trees, ſhrubs, and herbaceous plants; ranking,

ranking, however, with the latter, ſuch as had been called *Suffrutices*, or ſhrubby. *Trees* he divides into nine claſſes, accounting the laſt anomalous; *Shrubs* into ſix; and *Herbs* into forty-ſeven.

In the progreſs of his improvements afterwards, he reduced theſe claſſes to thirty-three. His method, which is indeed extremely elaborate, will beſt be ſeen by a view of the claſſes. It will, however, be but juſtice to refer the account to the laſt edition, that it may appear in the greateſt perfection which he gave it.

To this book Mr. RAY has ſubjoined a clear, conciſe view, and a ſynoptical table, of the ſyſtem of CÆSALPINE, and gives his reaſons for not adopting it; although he candidly confeſſes his obligations to the author, whom he acknowledges to be the parent of ſyſtem.

In 1683 and 1684, Mr. RAY and Dr. *Tancred* ROBINSON exchanged ſeveral letters, while the latter was on a foreign tour, relating to various undetermined facts in natural hiſtory; among which, it had been difficult to ſettle the exact ſpecies of the *Macruſe*,

Macruse, a bird allowed by the Roman Catholics to be eaten in Lent. Their observations relating to this particular were published in the *Phil. Trans.* for 1685, in No. 172. It proved to be the *Scoter*, or *Anas nigra Linnæi.*

We are now come to that performance, which LINNÆUS and HALLER so justly stile *Opus immensi laboris*; and which, considered as the work of one man, has perhaps been exceeded by none, unless indeed by that of *John* BAUHINE, who, however, did not live to put the finishing hand to his labour.

Mr. RAY informs us, that it was at the persuasion of his friend, Mr. WILLUGHBY, that he began to collect materials, with a view to a General History of Plants. But that, after the loss of his friend in 1672, he relaxed; and, on hearing that Dr. MORISON was employed on a similar design, from which considerable expectations were formed, at length gave up his purpose. On the decease of Dr. MORISON in 1683, who left the much greater part of his work unfinished, by the persuasion of his friends, and particularly of Mr. HOTTON, to whom it was dedicated, he resumed his design, and prosecuted the work with vigour. We cannot sufficiently admire the wonderful assiduity and address of this great man, which enabled him, in four years, to collect such a stock of matter, as to furnish two folio volumes, of near one thousand pages each. It even does not appear that he had the assistance of an amanuensis in this labour; which he effected, however, with a skill and judgment that gained him the applause of all succeeding masters in the science.

This important undertaking was intended by the author to comprehend the whole botany of the age, by describing separately, and reducing to his own system, all the plants of the BAUHINES, and of those who had enlarged the stock by subsequent discoveries. These, at the publication of RAY's first volume, were, the plants of *Mexico*, from HERNANDEZ; those of *Brasil*, from PISO and MARCGRAAVE; and of the *East Indies*, from BONTIUS. The rare plants of *Italy*, from ZANONI; the new plants of MORISON, BREYNIUS, and MENTZEL.

The

The *Sicilian* plants of BOCCONE; but above all, the vaſt treaſure of the ſix firſt volumes of the HORTUS MALABARICUS; with many from works of leſſer note.

After prefixing an inſtructive liſt of the writings of near an hundred botanical authors, quoted by him in the body of the book, and giving an explanation of terms, there follows a very comprehenſive account of the philoſophy of vegetables; in which the anatomy and phyſiology of plants, from MALPHIGI, from GREW, and from his own experiments; the differences of the parts of vegetables, from JUNGIUS and others, are explained and illuſtrated, with that judgment and knowledge of the ſubject, and with that conciſeneſs and methodical accuracy, which, I believe, had rarely, if ever, been equalled by preceding writers. This has rendered the introduction to his Hiſtory, a choice compendium of all that was valuable in the ſcience of his day; nor is the information it conveys ſo far ſuperſeded by any ſubſequent diſcoveries, as to render it, even now, an unintereſting tract. It is not eaſy to refer the modern ſtudent to a more perfect view of the state of this science near the close of the last century, than will here be found; while the work itself exhibits the great improvement it had received, since the beginning of the same period, and to which the author had himself contributed in an eminent degree.

The first volume was published in the year 1686, under the following title: "HISTORIA PLANTARUM GENERALIS: *species hactenus editas aliasque insuper multas noviter inventas et descriptas complectens; in qua agitur primò de plantis in genere, earumque partibus, accidentibus, et differentiis; deinde genera omnia tum summa tum subalterna ad species usque infimas, notis suis certis et characteristicis definita, methodo naturæ vestigiis insistente disponuntur; species singulæ accuratæ describuntur, obscura illustrantur, omissa suppleatur superflua resecantur, synonyma necessaria adjiciuntur: vires denique et usus recepti compendiò traduntur. Accesserunt Lexicon Botanicum, et Nomenclator Botanicus, cum indicibus necessariis nominum morborum et remediorum.*" Folio. Vol. I. pp. 984. Vol. II. pp. 985—1944. *preter indices.* 1688.

In the general arrangement of the ſubject, according to his own ſyſtem, he has in various inſtances improved the claſſes. At the head of each book or claſs is prefixed a ſynoptical table of all the chapters or *genera*.

In the particular diſpoſition, after prefixing to each chapter the etymology of the generical name, he gives the character of the *genus*; and in the enumeration of the ſpecies, quotes at length the ſynonyms of *Caſpar* BAUHINE, from his "*Pinax*," and thoſe of *John* BAUHINE, GERARD, and PARKINSON, from their reſpective hiſtories; ſeldom introducing others, where the plant was known to any of theſe writers.

His deſcriptions of the old plants are taken from the above-mentioned authors. They are commonly abridged, however; and in numberleſs inſtances amended, from his own knowledge of the plants. He fails not to notice from whom they are taken, and has every where diſtinguiſhed the *Britiſh* plants from the exotics. He has carefully marked all ſuch as he had not had an opportunity of inſpecting himſelf. He adds the places of growth, and times of flowering, and ſubjoins ſelect obſervations, from the moſt reſpectable authorities, relating to the qualities and various uſes of them.

In the "Hiſtory of Trees," the nobler and more capital parts of the vegetable kingdom, as being dignified by the variety of their uſes in human œconomy, he has extended his reſearches, and collected, with much aſſiduity, a greater variety of intereſting particulars. Mr. RAY has purpoſely avoided entering into nice and critical diſquiſitions relating to the ſpecies; for, beſides that this would have ſwelled his work to an enormous bulk, it was become leſs neceſſary, after the deſcriptions given by *John* BAUHINE, CLUSIUS, and others, ſo much ſuperior to thoſe of their predeceſſors; and the more curious and critical examiner might be referred to theſe authors, for ample ſcope to his curioſity.

Mr. RAY has deſcribed, in theſe volumes, about 6900 plants; including, however, in this number, many which modern botaniſts have ſince conſidered as varieties.

The

The *Addenda* to the second volume contain several interesting catalogues; such are those of ZANONI's History, consisting of new *Italian*, *Swiss*, and *Milanese* plants; those of BREYNIUS; a catalogue of the plants of *Virginia*, observed by Mr. BANISTER; and a compend of those of *Mexico*, from HERNANDEZ, who, at the expence of sixty thousand ducats, had procured the paintings of 1200 species, which perished in a fire of the Escurial.

In the preface to the first volume, Mr. RAY acknowledges his obligations for assistance received from many of his friends. Among those who had more essentially contributed to enrich his work, were, Sir *Edward* HULSE, Dr. *Tancred* ROBINSON, Dr. SLOANE, and his near neighbour, Mr. DALE. To these he adds, in the second volume, the names of *William* COURTINE, Esq; of the Middle Temple, Dr. PLUKENET, Mr. DOODY, and Mr. PETIVER.

There are copies of RAY's History, with the date of 1693; but I believe the title-page only to be new, the remaining copies of the impreſſion by *Faithorne*, falling into the hands of *Smith* and *Walford* about that time. Foreign writers mention an edition ſo late as 1716; but this I ſuſpect to be a miſtake, or owing to another transfer of the copies.

After the firſt edition of the "*Catalogus Plantarum Angliæ*" was out of print, Mr. RAY had been exhorted by his friend, Dr. *Ralph* JOHNSON, to arrange the ſecond according to ſyſtem; but not having ſufficiently elaborated his method, at that time, he declined it; and it came out in 1677, in the alphabetical order.

A third edition being wanted, however, after the publication of the "Hiſtory of "Plants," he meditated throwing it into the ſyſtematic form; and, in the mean time, put forth, in 1688, "FASCICULUS STIRPIUM BRITANNICARUM, *poſt editum Plantarum Angliæ Catalogum obſervatarum.*" Lond. 8°. By this little volume, a conſiderable acceſſion was made to *Engliſh* botany: ſeveral very rare mountainous or *Alpine* plants, from *Wales*; ſome ſcarce ones from

from *Cornwall*; ſea plants; new fungi; moſſes, and graſſes, make their firſt appearance in this little catalogue.

The "SYNOPSIS," although finiſhed for the preſs ſoon after this "*Faſciculus*," was not publiſhed, owing to the delay of the printer, till 1690, when it appeared under this title, "SYNOPSIS METHODICA STIRPIUM BRITANNICARUM, *in qua tum notæ generum characteriſticæ traduntur, tum ſpecies ſingulæ breviter deſcribuntur*: 250 *plus minus novæ ſpecies, partim ſuis locis inſerantur, partim in appendice ſeorſim exhibentur; cum indice et virium epitome.*" 8°. pp. 317.

As Mr. RAY had dedicated the "*Alphabetical Catalogue*" to his great friend and *Mecænas*, *Francis* WILLUGHBY, Eſq; ſo he now ſhews the ſame reſpect to *Thomas*, the only ſurviving ſon of his much-honoured patron; whom he exhorts to purſue the example of his excellent father, and for whom he pours forth, in the moſt energetic language, all thoſe ardent wiſhes which gratitude and reſpect for the memory of the father, and love for the pupil, could alone inſpire.

In

In the preface, Mr. RAY acknowledges the aſſiſtance he received from Mr. BOBART, ſuperintendant of the garden at *Oxford*; Mr. DALE, his neighbour, a learned and ingenious apothecary at *Braintree*; Mr. *Matthew* DODSWORTH; Mr. *Samuel* DOODY, an apothecary in *London*, memorable for having been the firſt who extended the *Cryptogamous* claſs; Mr. *Thomas* LAWSON, of *Strickland*, in *Weſtmorland*; Mr. *James* NEWTON, a diligent and ſkilful botaniſt; Dr. *Edward* LLOYD, of *Oxford*; Mr. *James* PETIVER; Dr. *Robert* PLOTT; Dr. PLUKENET; Dr. *Hans* SLOANE; Mr. *William* SHERARD, at that time fellow of St. *John's* College, *Oxford*; and Dr. *Tancred* ROBINSON, to whom Mr. RAY communicated his manuſcript of this work, and for whoſe corrections and additions, he held himſelf eminently obliged.

The *Appendix* contains a liſt of ſcarce plants, communicated by Mr. BOBART; ſome new plants by Mr. SHERARD; a liſt of thoſe of *Jerſey*, by the ſame; new and rare ſpecies, with critical obſervations, from Dr. PLUKENET; *muſci* and rare plants, by Mr.

Mr. DOODY; emendations and additions, by Dr. *Tancred* ROBINSON; and a catalogue of thirty-four ſpecies, common both to *England* and *Jamaica*, communicated by Dr. SLOANE. In this work, Mr. RAY has thrown the obſervations on the qualities and uſes into the index.

From this time the "SYNOPSIS" became the pocket companion of every *Engliſh* botaniſt. It contributed not a little, both to facilitate and improve the ſcience. It diffuſed the knowledge of ſyſtem; and, by obliging thoſe who wiſhed for improvement, to attend more minutely to generical characters, led to a nicer diſcrimination of both *genera* and *ſpecies*.

CHAP.

CHAP. 19.

Account of Ray's *works continued*—Wisdom of God manifested in the Works of the Creation—Physico-theology—Ray *considered as a zoologist*—*The first truly systematic writer on animals*—Synopsis Quadrupedum—Avium et Piscium—*Publishes* Rauwolf's Travels, *with valuable additions*—Stirpium Europæarum Sylloge—*Controversy with* Rivinus—*Provincial catalogues of plants for* Gibson's Camden—*Great improvement to* English *Botany given by* Ray—*Evidenced by the second edition of the* Synopsis—De variis Plantarum Methodis—Epistola ad Rivinum—*His* Persuasive to a Holy Life.

RAY.

TO this period Mr. RAY had appeared to the public principally as a naturalist; but he now united to this character that of the theologist. It is needless to say, that he succeeded in this department, perhaps beyond most of those who had before written on the same subject. His first publication of this kind, we are told, was originally,

originally, and in its outlines, *College Exercises* only, or *Common Places*. These he now wrought up, and enlarged into a convenient volume, and trusted it to the care of his friend, Dr. *Tancred* ROBINSON, who procured five hundred copies to be printed, under the following title: "THE WISDOM OF GOD MANIFESTED IN THE WORKS OF THE CREATION." 8°. 1691. It was reprinted the next year. The eleventh edition was published in 1743; and a twelfth in 1758; and, I believe, several times since: and it has been translated into foreign languages. These are sufficient testimonies of the esteem with which it was received by the public.

It is not immediately within my plan to enlarge on this work, or to determine whether the arguments *à priori*, or *à posteriori*, are best calculated to obtain the object of it, "Demonstration of the Being of a God." "*Qui historiam naturæ, naturæ etiam Creatorem colit.*" I may be allowed to observe, that Mr. RAY, from that comprehensive view of nature which his mind embraced, was singularly well qualified to display the manifold wonders

wonders of the creation, and the wiſdom of its omnipotent Author. And thus, while his penetrating views enabled him to unfold the various œconomy and evolutions of nature to the greateſt advantage, his piety and humility give a force to his reaſonings and deductions, that carries with it a conviction of that great truth he ſo ſincerely wiſhed to inculcate.

The favourable acceptance the public gave to the "Demonſtration," encouraged Mr. RAY to publiſh, the next year, his "THREE PHYSICO-THEOLOGICAL DISCOURSES concerning the primitive Chaos, "and Creation of the World. The general "Deluge, its cauſes and effects. The Diſ-"ſolution of the World, and future Confla-"gration." 8°. 1692. and 1693. 1713. pp. 456. 1721. 1732. It is embelliſhed with a plate of the *Apamæan* medal, and three tables of figured foſſils; and is dedicated to Archbiſhop TILLOTSON.

This work is a convincing proof of the extenſive reading, the various erudition, and multifarious knowledge, of this great and good man. Independent of all the theories

it

it contains, this volume exhibits ſuch an aſſemblage of facts, relating to the ſtructure of this globe, to the changes it has undergone, and to the hiſtory of figured foſſils, that it may be read to advantage, even in this age of advanced curioſity, and knowledge in the profeſſed object of this book. Even the faſtidious critic, who is verſed in all the more modern theories, down to the " Epochas of Nature," and thoſe of *M. De* LUC, and *De* SOULAVIE, will allow that this volume, when reſpect is had to the time of its publication, muſt have conveyed a large ſhare of intelligence to thoſe who were capable of gratification from diſquiſitions of this nature; and that, with a deference to the opinions of the day, there is yet a freedom of enquiry that diſtinguiſhes the author, as a friend to true philoſophy, and as a modeſt and candid enquirer after truth, in thoſe points of natural hiſtory, which ſtill continue, and probably long will, to be involved in great obſcurity.

In this year, Mr. RAY wrote ſome " Obſervations on the Planting of *Maize* " inſtead of *Peaſe*," occaſioned by a pro-

 poſal

poſal of Sir *Richard* BULKLEY, in which he ſays, that he had found the greateſt yield of peaſe to be twenty barrels reaped for one ſown; whereas, from one grain of *Indian wheat*, he had calculated the produce would be upwards of 2000 grains for one. Theſe Obſervations were printed in the *Phil. Tranſ.* N° 205. Mr. RAY was not ſanguine in his expectations from the culture of that grain; neither have ſubſequent trials proved the utility of it in this climate.

The botanical labours of this eminent man were now remitted, at leaſt for ſome time; and we find, that after the publication of his "*Hiſtory*," and the "*Synopſis*," his exertions were turned into another channel, in which he alſo ſtood unrivalled in his day. It was not botany alone that he raiſed from a drooping ſtate; to zoology, conſidered as a ſcience, he might be ſaid to have given birth, in theſe kingdoms; ſince, except what himſelf and Mr. WILLUGHBY had performed, nothing of importance on the hiſtory of animals exiſted.

TOPSELL's "Abridgment of GESNER," MOFFAT's

MOFFAT's "Book on Insects;" and the short and imperfect essays of CHARLETON, in his "*Onomasticon*," and of MERRET, in his "*Pinax*," were almost the only *English* writers to be consulted. To assert that better helps were wanted, is not to injure, or to degrade those authors. Mr. RAY had been urged by his friends, and particularly by Dr. ROBINSON, to undertake an entire FAUNA ANGLICA, and a history of *Fossils* also; but age and infirmities began now to oppress him, and he thought himself inadequate to the attempt. He lived, however, to perform more than his fears, or his humility permitted him to hope.

In 1693, he published his "SYNOPSIS METHODICA ANIMALIUM, QUADRUPEDUM, *et* SERPENTINI GENERIS; *vulgarium notas characteristicas, rariorum descriptiones integras, exhibens: cum historiis et observationibus anatomicis, perquam curiosis. Præmittuntur nonnulla de animalium in genere, sensu, generatione, divisione, &c.*" Lond. 8°. pp. 336.

In this volume we see the first truly systematic arrangement of animals, since the days of ARISTOTLE; an arrangement which his ſucceſſors in the ſame line have equally applauded, and availed themſelves of. It is profeſſedly the baſis of that method, by which the preſent eminent zoologiſt of this nation, has choſen to convey his learned publications, and by which he has not leſs happily diffuſed a taſte for this ſcience, than he has ſucceſsfully improved its ſtore.

In treating on animals in general, introductory to his work, he diſcuſſes ſome important queſtions, which had not then ceaſed to agitate the philoſophical world. He controverts, with extreme force of reaſoning, the ideas of equivocal or ſpontaneous generation; the *Lewenhoekian* hypotheſis; and that of all animals being created from eternity, and only latent in an involved ſtate. I know not where the reader can ſee theſe queſtions diſcuſſed with equal conciſeneſs and judgment united.

Mr. RAY's Diſtribution of Animals is not wholly founded, as to the grand diviſions, on the *Ariſtotelian* diſtinctions; though he admits many of them. It is not within my plan to enter on this ſubject; it is ſufficient to obſerve, that *Quadrupeds* here form two great diviſions,

divisions, as they are *hoofed* or *digitated*; the former, as they are *whole* or cloven; the latter, as they are *divided* into more, or fewer claws; admitting also of subdivisions or *genera*, from the number of the claws, and in some, from the consideration of the teeth.

At the time when Mr. RAY lived, few people had acquired a taste for this kind of knowledge, and commerce had not lent her friendly aid, as in later times. What animals came under his own inspection, he has described with his accustomed accuracy; from GESNER and ALDROVAND he borrows his descriptions of others; and many later discovered subjects he drew from PISO and MARCGRAAVE, from CLUSIUS, HERNANDEZ, LAET, and NIEREMBERG.

In the course of this work, he has, in various instances, given the anatomical structure, from Dr. TYSON, from the "*Parisian* "Dissections," and other works. Throughout the whole, he has shewn how intimately he was acquainted with the learning of the ancients, and particularly with ARISTOTLE, whom, as the parent of zoological knowledge, he failed not to consult on all occasions, but by no means implicitly to follow, in his subtleties and obscurities.

On finishing the "Synopsis of Quadrupeds," Mr. RAY immediately drew up that of the *Birds* and *Fishes*. This was an easier task at this time, since they are to be considered as compends of his preceding labours with his friend, Mr. WILLUGHBY; although there were many things new in both, and that of the Fishes was very greatly improved as to the arrangement and method. He informs us, that the additions were, the *Mexican* birds, from HERNANDEZ; some descriptions of new species, out of NIEUHOFF; MARTIN's Birds and Fishes of *Greenland*; SIBBALD's Whales; SLOANE's *Jamaica* Birds and Fishes; and some from the *Leyden* Catalogue, by Dr. ROBINSON.

In these branches of nature, Mr. RAY again appears as the parent of method. The accurate BRISSON regards RAY and WILLUGHBY, as the first true systematic writers on birds. These works were finished

in

in the year 1693 or 1694, as we learn from Mr. RAY's letters, and from the teſtimony of his friend and neighbour, Mr. DALE. Yet, excellent as they were, ſo ſcanty was the taſte for natural hiſtory at this period, that the manuſcripts lay unpubliſhed in the bookſellers hands, till they were purchaſed by Mr. INNYS, and prepared for the preſs by Dr. DERHAM, who added the figures, and inſerted Mr. BUCKLEY's Birds from *Madraſs*, and Mr. JAGO's *Corniſh* Fiſhes. They were publiſhed in 1713, under the titles of "SYNOPSIS METHODICA AVIUM." 8. pp. 198. t. 2; and "SYNOPSIS METHODICA PISCIUM." 8°. pp. 162. t. 1.

In the ſame year, 1693, Mr. RAY became the editor of a tranſlation of "Dr. RAUWOLF's Travels." This phyſician, who was the next after BELON, whom the love of natural hiſtory alone, led to travel into the eaſt, ſpent the years 1573-4-5 in traverſing *Syria*, *Meſopotamia*, *Paleſtine*, and *Ægypt*, induced, as he tells us, by his deſire to behold, in the native places, the plants of the *Greek* and *Arabian* phyſicians.

His "Travels" having been publiſhed in 1583, in the *German* language, had hitherto been locked up from the *Engliſh* reader. Sir *Hans* SLOANE having, however, read them, was induced, in concert with Capt. HATTON, to procure a tranſlation of them, which was done by *Nicholas* STAPHORST. This verſion was put into Mr. RAY's hands, to reviſe and correct. He did more; he made a choice ſelection from other authors, who had made the ſame tour, BELON, ALPINUS, Sir *George* WHELER, &c. and he drew up a Catalogue of the more rare Plants of thoſe countries through which RAUWOLF travelled; and added liſts of thoſe of *Ægypt* and *Crete*. From this circumſtance, the book has gone by the name of "RAY's "COLLECTION OF TRAVELS;" and it was reprinted with his own "Obſerva-"tions," in 1738. RAUWOLF made an "*Herbarium*," while in the eaſt; which, with his *European* plants, conſtituted four large volumes. Theſe became the property of Queen *Chriſtina*, and afterwards, by her means probably, of *Iſaac* VOSSIUS, who informed

formed Capt. HATTON, that 400 *l.* ſterling had been offered for them. They were purchaſed of his heirs by the univerſity of *Leyden*; and the late Dr. *Frederick* GRONOVIUS conſtructed from them an elegant and learned "*Flora Orientalis*;" of which he much enhanced the value, by prefixing to it *Melchier* ADAMS's "Life of RAU- "WOLF," with large additions of his own.

The "CATALOGUS STIRPIUM IN EXTERIS REGIONIBUS OBSERVATARUM" being out of print, Mr. RAY was induced to give a new edition of it, with ſuch large augmentations, as to make it a new work. He added from CLUSIUS, from the BAUHINES, and other authors, a number of plants growing in thoſe regions through which he paſſed in his tour; and ſo many catalogues from other authors, as to render it a tolerably complete liſt of all the *European* plants, not natives of *England*. As it does not immediately reſpect *Engliſh* botany, it will be ſufficient to recite the title-page, from which its ſcope may be underſtood:

"STIRPIUM EUROPÆARUM *extra Britannias*

tannias naſcentium SYLLOGE. *Quas partim obſervavit ipſe, partim à C. Cluſii Hiſtoria; C. Bauhini Prodromo, et Catalogo Baſilienſi; F. Columnæ Ecphraſi; Catalogis Hollandicarum A. Commelini; Allorſinarum M. Hoffmanni; Sicularum P. Bocconi; Monſpelienſium P. Magnoli; collegit J.* RAIUS. *Adjiciuntur Catalogi rariorum Alpinarum et Pyrenaicarum, Baldenſium, Hiſpanicarum Griſleii, Græcarum et Orientalium, Creticarum, Ægyptiacarum, aliique: ab eodem.*" Lond. 1694. 8°. pp. 445.

In the preface to this work, Mr. RAY, for the firſt time, entered into controverſy; having taken occaſion to throw out ſome ſtrictures on the method of botany publiſhed by RIVINUS in 1690. It is not enough intereſting at this day to dwell on the nature of it. It is ſufficient to obſerve, that our veteran in ſcience was diſſatisfied with the *German*, for throwing the trees promiſcuouſly into the claſſes with other plants, and for breaking into the natural orders, for the ſake of agreement in the flower alone. In fact, RIVINUS's method being founded wholly on the flower, to which

which part RAY had paid but ſmall regard, the ſources of controverſy were endleſs; ſince the fundamental principles of each were totally irreconcileable.

About this time Mr. RAY communicated "The Provincial Catalogues of Plants," printed at the end of each county, in the edition of "CAMBDEN'S *Britannia*," publiſhed in 1695 by Mr. GIBSON. His repeated travels throughout moſt parts of *England*, for the ſole purpoſe of inveſtigating the ſubjects of nature, had enabled him to accompliſh more than had been done by any man before his time; and his unqueſtionable ſkill and accuracy, added an authenticity to theſe liſts, which could not eaſily have been derived from any other hand.

To the county of *Cornwall* Mr. RAY added many other particulars; which, however, were not printed, probably becauſe the correſponding circumſtances could not be procured from other counties. Theſe were, "Catalogues of the Sea Fiſh, and Sea Fowl, with the ſynonyms;" ſome account of two or three ſorts of ſtone dug there; of ſea ſand, as manure; an account of the *hurlers*,

hurlers, and other ſtones; and notices reſpecting the manners and language of the inhabitants.

Such as are converſant with that ſcience, which was the favourite object of Mr. RAY, muſt be ſenſible that nothing could have happened more conducive to the revival and improvement of it at this juncture, than the circumſtance of its having been taken up by a man of ſuch patient induſtry, capable at the ſame time of giving it all the embelliſhments, and advantages that learning could afford. They will readily grant that his writings and example alone, added more vigour, and brought more diſciples to this ſchool of natural ſcience in *England*, than all the exertions of foregoing writers.

I cannot confirm and illuſtrate the truth of this poſition more effectually, than by calling to the attention of the curious in this kind of knowledge, the vaſt augmentation it acquired, in the interval between the publication of Mr. RAY's "*Catalogus Plantarum Angliæ*," and that of the "*Synopſis*;" and more eſpecially between the time of the firſt and ſecond edition of the latter

latter work; during which, exclufive of the difcovery of many fubjects, among what were called the *more perfect* plants, a new and very extenfive field had been opened, by exciting attention to the *lefs perfect* (as they were then accounted) and minuter kinds of vegetables, the *Fungi*, *Fuci*, *Mufci*, and *Algæ*, known now by the name of *Cryptogamiæ*. During the firft of thefe periods, 250 fpecies had been added to the *Englifh Flora*; and the acceffion in the laft exceeded that number.

In no part of *Europe* had the fame progrefs been made in the inveftigation of thefe hitherto much-neglected fubjects, as in *England*, during the period above mentioned. This is fufficiently evinced by comparing the fecond edition of the "SYNOPSIS" with the contemporary writings of foreign botanifts.

This fecond edition of the "SYNOPSIS" was printed in 1696. 8°. pp. 346. Mr. RAY himfelf had but a fmall fhare in the augmentations that were made to this edition. His advancing years and infirmities prevented him from making excurfions. His principal

principal auxiliaries are mentioned in the preface; in which, additional to the names in the former "SYNOPSIS," we meet with those of Mr. *Edward* LLHWYD, *Walter* MOYLE, Esq; and Mr. *William* VERNON, fellow of St. *Peter's* College, *Cambridge*.

To those who are sensible of the obligations which the science owes to Mr. RAY, it cannot but be grateful to read, with what satisfaction the good man records, in this preface, the progress he had lived to see his favourite study make in his own country, and with what delight he augurs and contemplates its future improvement. In the space of little more than twenty years, and under his own pen, he had seen the *English Flora* acquire an accession of upwards of 500 new subjects. The "CATALOGUS PLANTARUM ANGLIÆ" of 1670, containing about 1050, and the second edition of the "SYNOPSIS" in 1696, full 1600 species; and, notwithstanding these have not all stood the test of the discriminating character of the present age, yet, in justice to this great man, and his associates, it must be acknowledged,

 that

that the retrenchments are comparatively few.

With this edition of the "SYNOPSIS," was publiſhed the "DISSERTATIO DE VARIIS PLANTARUM METHODIS BREVIS;" in which Mr. RAY ſhews, that the ſeparation of plants into claſſes and *genera* from the fructification alone, muſt be a very gradual and progreſſive affair; that it was not eaſy to exclude the habit from having a ſhare in this diſtribution, ſince there were many plants that were ſeldom or never ſeen in flower by the early botaniſts. He alſo obſerves, that numberleſs plants, which agree in the ſtructure of the flower; differ materially in habit, and others *vice verſa.* And although his own method is principally founded on the fruit, yet he freely acknowledges its imperfections; but thinks the ſame objections hold againſt the flower; which he illuſtrates by ſhewing, in TOURNEFORT's ſyſtem, the uncertainty of the bounds between the *Flores infundibuliformes, hypocrateriformes,* and the *caryophyllei.* If Mr. RAY paid leſs regard to the flower than its importance ſeemed to demand, it ſeems rather to have ariſen from the

the principles of his method, than from his want of opportunities of examination, owing to his distance from botanical gardens, as was alledged by his opponents; a circumstance, however, which he very feelingly laments in the preface to his "METHODUS," and elsewhere.

To this is annexed, "EPISTOLA *de* METHODO PLANTARUM *viri clarissimi D. A. Q. Rivini ad Raium, cum ejusdem responsoria, in qua D. Jos. Pitton Tournefortii, M. D. Elementa Botanica tanguntur.*"

On the method of RIVINUS, Mr. RAY, as was before noticed, had thrown out some strictures in the preface to his "*Sylloge,*" which drew from that author the answer here published, and Mr. RAY's reply; in which our author takes occasion also to defend his method from the objections of TOURNEFORT, who had been unbecomingly severe in some animadversions made in the "*Elements of Botany,*" published in 1694. TOURNEFORT, however, afterwards did ample justice to the merits of our author.

The modern botanist sees that all these controversies are become too little interesting

ing to dwell upon at this time. The principles of the *Corolliſtæ*, and the *Fructiſtæ*, as LINNÆUS ſtyles them, can never be aſſimilated, and all attempts to reduce the whole vegetable kingdom into natural claſſes have hitherto failed.

In 1697, he wrote "Some Obſervations "on the Poiſonous Effects of a Root eaten "inſtead of Parſneps," ſuppoſed to have been that of the Hemlock; but of which Mr. RAY had ſome doubt, alledging, that it was more probably the *Cicutaria vulgaris*, (*Chærophyllum ſylveſtre*, Lin.) See *Phil. Tranſ.* N° 231. In N° 238, he communicated "Remarks on the Poiſonous Effects "of the *Oenanthe crocata*," too fatally confirmed by later miſtakes of the ſame kind.

In the year 1700, Mr. RAY publiſhed "A PERSUASIVE TO A HOLY LIFE, from "the Happineſs which attends it both in "this World and in the World to come." Lond. 8°. Reprinted in 1719. pp. 126. He tells us it was drawn up at the requeſt of his friend, Mr. *Edmund Elys*, and that it is compoſed on the model of Biſhop WILKINS's

KINS's "Treatiſe on Natural Religion." It is wholly of a moral and practical nature, written in a plain, but forcible and argumentative ſtyle, and is entirely deſtitute of any of thoſe enthuſiaſtic or myſtical opinions, which ſo highly tinctured the writings of many divines of the laſt century. On the contrary, Mr. RAY, ever conſiſtent and rational, although he deduces his principal motives to the practice of virtue, as conducive to happineſs, even in this life, from the precepts of Chriſtianity; yet does not diſdain, particularly in treating on pleaſure, on riches, and the advantages of temperance, to enforce his arguments by opinions and apophthegms from the writings of the philoſophers and moraliſts of ancient *Greece*, and *Rome*.

CHAP.

CHAP. 20.

Account of Ray *continued—Improved edition of the* Methodus Plantarum—*Outlines of* Ray's *system—Third volume of the* Historia Plantarum —Methodus Insectorum—*His Death and Character.*

RAY.

THE peaceable mind of Mr. RAY could not delight in the contentious field of controversy; on the contrary, he regretted the occasions that drew him into it: yet were they not without use, since they unquestionably stimulated him to purify and correct his own *Methodus*. This he effected in the year 1698, although at this time much declined in his health, being afflicted with ulcers of the lower extremities, the pain of which rendered his nights frequently sleepless, and wholly prevented him from making excursions to *London*, as he much desired, to examine the gardens and *herbaria* of the curious.

So ſmall, however, was the demand for books in this ſcience, at the entrance of this century, that the *London* bookſellers were unwilling to riſk the printing of it: and it was finally ſent to *Holland*, and printed at *Amſterdam*, under the care of Dr. HOTTON, the botanical profeſſor at *Leyden*, who ſuperviſed the preſs, and procured 1100 copies to be thrown off, under the title of "METHODUS PLANTARUM EMENDATA ET AUCTA: *accedit Methodus Graminum, Juncorum, et Cyperorum ſpecialis.*" 8°. pp. 202. 1703. Dr. HOTTON gave a further ſanction to the ſyſtem of his friend; he taught it in his lectures to the pupils of that univerſity, and informed Mr. RAY of the good acceptance it met with on the continent, particularly in *Italy*. This volume was reprinted at *Amſterdam* in 1710, and at *Tubingen* in 1733.

In the preface he recapitulates his own progreſs in the formation of his ſyſtem, and dates it from the tables drawn up in 1667 for the uſe of Biſhop WILKINS. He very juſtly reprehends Dr. MORISON, for affecting to have formed his method entirely from his own obſervations, without acknowledging the ſmalleſt aid from former writers; briefly recites his objections to the methods of RIVINUS, TOURNEFORT, and HERMAN; and defends his own. He eſtabliſhes ſome axioms, to be obſerved in framing a ſyſtem of botany. In fact, Mr. RAY's method, though he aſſumes the fruit as the foundation, is an elaborate attempt, for that time, to fix natural claſſes. He eſtabliſhes it as a rule, that no plant is to be ſeparated from its tribe for a ſingle note of difference; but that all are to be aſſimilated, as far as habit will allow. The characters of the *genera* are, however, highly incongruous; they are taken from vague principles, ſuch as the ſhape of the leaf, colour of the flower, taſte, ſmell, and ſometimes from the ſize of the plant, and other as unſtable diſtinctions.

In this amended edition, Mr. RAY ſtill adheres to the ancient diviſion into trees and herbaceous plants, having dropped the diſtinction of ſhrubs, preſerved in the firſt edition. Here, all herbaceous, and ſhrubby-ſtalked plants are divided into twenty-five *genera* or claſſes; as follow:

1. *Submarinæ.*
2. *Fungi.* In the firſt *Methodus*, theſe two claſſes were formed into one claſs, or ſynoptical table.
3. *Muſci.*
4. *Capillares.*
5. *Apetalæ.* Before, in two tables.
6. *Planipetalæ lacteſcentes.*
7. *Diſcoideæ.* Before, in two claſſes.
8. *Corymbiferæ.* Before, in two claſſes.
9. *Capitatæ.* Before, in two claſſes.
10. *Herbæ ſemine nudo ſolitario, flore ſimplici perfecto.*
11. *Umbelliferæ.*
12. *Stellatæ.*
13. *Aſperifoliæ.*
14. *Verticillatæ.* Before, divided into two; *Herbaceæ, et Fruticoſæ.*
15. *Polyſpermæ.* Formerly, in two claſſes.
16. *Pomiferæ.*
17. *Bacciferæ.*
18. *Multiſiliquæ.*
19. *Vaſculiferæ, Monopetalæ.* Before, in three claſſes; *et Dipetalæ.*
20. *Siliquoſæ, et Siliculoſæ.* Formerly, in three claſſes; *et Anomalæ.*

21. *Pa-*

21. *Papilionaceæ; ſ. Leguminoſæ.* Formerly, in four claſſes.
22. *Pentapetalæ.* Before, in two claſſes.
23. *Floriferæ, Graminifoliæ.* Formerly, in four claſſes; *et Bulboſis affines.*
24. *Stamineæ, Graminifoliæ.* Before, in three tables.
25. *Anomalæ.*

Trees, and Shrubs.

26. *Arundinaceæ.*
27. *Flore a fructu remoto; ſeu Apetalæ.*
28. *Fructu umbilicato; ſ. Pomiferæ, et Bacciferæ.*
29. *Fructu non umbilicato; ſ. Pruniferæ.*
30. *Fructu ſicco; non ſiliquoſo, nec umbilicato; et Miſcellaneæ.*
31. *Siliquoſæ, non Papilionaceæ.*
32. *Siliquoſæ, Papilionaceæ.*
33. *Anomalæ.*

At this time, the conſideration of Mr. Ray's method is a matter of mere curioſity; yet, in juſtice to this great man, it muſt be remarked, that his ſyſtem, though leſs artificial than that of Cæsalpine, is much more highly elaborated than that of MORISON: and, though Mr. RAY muſt have taken infinite pains with it, yet is it difficult in practice; ſince the baſes of the claſſes are not uniform. Of the thirty-three, however, twelve are nearly compoſed of natural orders. Such are the following:

Fungi,	*Aſperifoliæ,*
Muſci,	*Verticillatæ,*
Capillares,	*Pomiferæ,*
Planipetalæ,	*Siliquoſæ,*
Umbelliferæ,	*Leguminoſæ,*
Stellatæ,	*Culmiferæ.*

The remaining claſſes are combined of ſubjects leſs connected by habit and ſtructure; and are therefore ſubject to more arbitrary rules, drawn from the conſideration of ſome one, or more parts, in the fructification.

In the "METHODUS *Graminum, Juncorum, et Cyperorum ſpecialis,*" annexed to this book, Mr. RAY's diſtribution reſts principally on what may be called the habit of the fructification; all thoſe *genera,* which in in the *Linnæan* ſyſtem are known by the names of *Phalaris, Alopecurus, Dactylis, Agroſtis, Aia, Poa, Briza,* &c. being called *Gramen* ſimply, with the epithets of the old authors annexed, expreſſive of the mode of bearing the parts of the fructification, whether in ſpikes, or panicles; as, *Gramen triticeum*; *Gramen loliaceum*; *typhinum*; *Gramen paniculatum*; *miliaceum,* &c. In this *Conſpectus,* however, all the ſpecies are introduced, to the amount of two hundred.

Sixteen years had now elapſed ſince the publication of his "Hiſtory of Plants;" in which interval botany had aſſumed a new face, and experienced a much greater revolution and acceſſion, than had ever taken place before. Syſtem had been ſtudied, and in ſome meaſure eſtabliſhed, both at home and abroad. An incredible number of new plants had been introduced, from all parts of the world, and cultivated with extreme care in the gardens of *Europe.* In the mean time, theſe circumſtances had given riſe to a great number of valuable publications. The remaining ſix volumes of that ineſtimable work, the "HORTUS MALABARICUS," had appeared: BREYNIUS, HERMAN, TOURNEFORT, PLUMIER, PLUKENET, BOCCONE, COMMELINE, BOBART, CUPANI, VOLKAMER, and RIVINUS, had enriched botany with valuable performances. Theſe large augmentations to the ſcience induced Mr. RAY, notwithſtanding his advanced years and ill health, to attempt a collection of theſe ſcattered materials, in order to form a ſupplemental volume to his "Hiſtory;" and his induſtry enabled him to effect his purpoſe. Additional to the aſſiſtances derived from all theſe printed works, he had acceſs, by the favour of Sir *Hans* SLOANE, to the MS. of his "Hiſtory of "*Jamaica* Plants" (of which the "*Prodromus*" had been publiſhed in 1696) with liberty to ſelect what he thought proper to his deſign.

From the ſame gentleman he enjoyed the benefit of an *Herbarium* of ſeveral hundred new and undeſcribed plants, collected in *Maryland*, by Mr. VERNON and Mr. KREIG, who had made a voyage thither for the

ſole

ſole purpoſe of gratifying their taſte in botany. Mr. PETIVER freely communicated his ſtores, at that time very ample, though afterwards abundantly more ſo; and Dr. SHERARD engaged, beſides ſupplying more than a thouſand ſpecies himſelf, to take the trouble of inſpecting the whole work before it went to the preſs, and of making ſuch corrections and additions as he judged proper.

It was the laſt of his works publiſhed in his life-time, and came out in 1704, with the following title:

" HISTORIÆ PLANTARUM TOMUS TERTIUS, *qui eſt* SUPPLEMENTUM *duorum præcedentium; ſpecies omnes, vel omiſſas, vel poſt volumina illa evulgata editas, præter innumeras fere novas et indictas ab amicis communicatas, complectens: cum ſynonymis neceſſariis, et uſibus in cibo, medicina, et mechanicis.*" Lond. folio. pp. 666; and the " *Dendrologia,*" pp. 135. App. pp. 137.

The diſtribution is the ſame as that of the two former volumes. In a compilation of this kind, collected from ſo numerous a ſet of authors, and in many inſtances from dried and imperfect ſpecimens, there muſt neceſſarily

neceſſarily ariſe a multitude of repetitions. The author was ſufficiently aware of this; but it was unavoidable. In this volume there are upwards of 11,700 plants enumerated.

The Appendix contains ſeveral catalogues, which muſt have been intereſting to the curious at that time. Father CAMELL, a learned Jeſuit of *Manila*, who had not only deſcribed, but delineated, a great number of the plants of *Luzone*, tranſmitted his work to Mr. RAY; and it forms an extenſive part of this Appendix. It muſt have been much regretted, that the Rev. Father had not been furniſhed with books to have enabled him to adapt the ſynonyms; ſince there are few inſtances in which any other names occur, than the *Spaniſh*, and the indigenous appellations of the natives and *Malays*.

Mr. RAY then gives a liſt of TOURNEFORT's oriental diſcoveries, from the "*Corollarium*;" thoſe of DAMPIER, from *New Holland* and elſewhere, and of MARTENS's *Greenland* Plants; of COMMELINE's Rare Exotics; a copious Catalogue of *Chineſe*, *Madraſs*,

Madrafs, and *African* Plants, communicated by Mr. PETIVER, of which, thofe from *Madrafs* had been collected by Mr. BROWNE, a furgeon at that fettlement; and laftly, a lift of the new, or hitherto very imperfectly defcribed fpecies, contained in Mr. PETIVER's *Hortus Siccus*, amounting to upwards of 800.

An advertifement had been printed at the end of the firft volume of Mr. RAY's "Hiftory," in 1688, inviting to a fubfcription for a fet of figures to the work; and it was propofed, that thofe belonging to each tribe or clafs, fhould be publifhed in regular fucceffion; but it did not fucceed. The fcheme was again revived, while the Supplement was printing; and, among other of Mr. RAY's friends, Dr. COMPTON, bifhop of *London*, had given his patronage, and ftrongly recommended it. Conferences were held with Dr. SHERARD, Sir *Hans* SLOANE, Dr. ROBINSON, and Mr. PETIVER, relating to it; but it was relinquifhed as impracticable.

Mr. RAY's infirmities were very preffing upon him during the later years of his life. In

In a letter, written in the ſpring of 1702, he informs Mr. DERHAM that he had not been half a mile from his own houſe for four years. Yet, under theſe circumſtances, he wrote his ſupplemental volume to his "Hiſtory of Plants," which, he ſays, had engroſſed almoſt his whole time for two years.

We have now brought Mr. RAY's botanical works to a concluſion; but his labours did not ceaſe here. His active and indefatigable mind prompted him, at the age of ſeventy-five, to begin a work on *Inſects*; to which he had been encouraged by Dr. DERHAM; and for which he had been accumulating materials during many years. This was intended to comprehend only the *Engliſh* ſpecies; although, at the ſame time, his friends were wiſhing to engage him to deſcribe the exotics of the *London Muſea*, which were then beginning to abound in theſe ſubjects.

He had paid ſome attention to the hiſtory of *Spiders*, indeed, many years before, when intimately connected with Dr. LISTER;

LISTER; but the greater part of his work was drawn up from his own actual descriptions, and partly from Mr. WILLUGHBY's papers, and the contributions of friends, Mr. PETIVER, Mr. DANDRIDGE, Dr. SLOANE, Mr. MORTON, and Mr. STONEFLEET.

He tells us, that in the later years of his life he had discovered 300 kinds of *Papilios*, diurnal and nocturnal; and knew there were many more. The *Beetles*, he observes, were as numerous, and the *Flies* not less so. I mention these circumstances to prove the extensive knowledge of nature which this extraordinary man possessed, at an æra when he stood so nearly alone in these branches of science. He did not live to finish this work. It was published by Dr. DERHAM in 1710, in 4°. pp. 398.

I believe Mr. RAY was the first who gave to these minuter animals a real and scientific distribution. He had drawn up a short "METHODUS INSECTORUM," which was published the year after his death. Of the history itself, it is sufficient to say, that it bears all the characters of that accurate, discriminating,

diſcriminating, and ſyſtematic genius, which guided him in all his reſearches in the field of nature; and that it is every where quoted by the eminent *Swede* with the higheſt commendations, for the faithful deſcriptions it contains.

Mr. RAY's infirmities and afflictions, painful and grievous as they were, did not, we are told, prevent him from proſecuting his ſtudies till within about three months before his death; which event took place on Jan. 17, 1704-5.

He died at *Black Notley*, and was buried, as Dr. DERHAM ſays, according to his own deſire, in the church of that pariſh. The writers of the "General Dictionary," in the mean time, inform us, that, "although the "rector of the pariſh offered him a place of "interment in the chancel of the church, "yet he modeſtly refuſed it, chooſing rather "to be buried in the church-yard with his "anceſtors, where a monument was erected "to him," as Dr. DERHAM relates, at the charge of ſome of his friends, with a *Latin* inſcription; which may be ſeen in the "General Dictionary," and in Mr. SCOTT's

"Remains;'

"Remains;" and of which I inſert a copy below *.

As Mr. RAY did not inherit any paternal eſtate, and had often refuſed preferment, his circumſtances could never have been affluent; and the legacy of Mr. WILLUGHBY is ſaid to have been the greateſt part of what he enjoyed. His own eſtate, whatever that might be, he ſettled on his wife. He

* The Inſcription on Mr. RAY's Monument.

Eruditiſſimi Viri JOHANNIS RAII, M.A.
Quicquid mortale fuit
Hoc in anguſto Tumulo reconditum eſt,
At ſcripta
Non unica continet Regio:
Et Fama undiquaque celeberrima
Vetat Mori.
Collegii SS. Trinitatis Cantab. fuit olim Socius,
Nec non Societatis Regiæ apud Londinenſes Sodalis,
Egregium utriuſque Ornamentum.
In omni Scientiarum Genere,
Tam divinarum quam humanarum
Verſatiſſimus:
Et ſicut alter Solomon (cui forſan unico ſecundus)
A Cedro ad Hyſſopum,
Ab Animalium maximis ad minima uſque Inſecta
Exquiſitam nactus eſt Notitiam.

Nec

He had four daughters, three of whom ſurvived him. " He left a ſmall legacy to the " poor of his own pariſh, and five pounds to " Trinity College, in *Cambridge*, to pur- " chaſe books for the library there. All " his collections of natural curioſities he " beſtowed

Nec de ſtantis ſolum quæ patet Terræ Facie,
Accuratiſſimè diſſeruit;
Sed et intima ipſius Viſcera ſagaciſſimè rimatus,
Quicquid notatu dignum in Univerſi Naturâ
Deſcripſit.
Apud exteras Gentes agens,
Quæ aliorum Oculos fugerant, diligenter exploravit,
Multaque ſcitu digniſſima primus in Lucem protulit.
Quod ſupereſt, eâ Morum Simplicitate præditus,
Ut fuerit abſque Invidiâ doctus:
Sublimis Ingenii,
Et (quod raro accidit) demiſſi ſimul Animi et modeſti.
Non Sanguine et Genere inſignis,
Sed (quod majus)
Propriâ Virtute illuſtris.
De Opibus Tituliſque obtinendis
Parum ſollicitus,
Hæc potius mereri voluit, quam adipiſci:
Dum ſub privato Lare ſua Sorte contentus,
Fortunâ lautiori dignus conſenuit.
In Rebus aliis ſibi Modum facilè impoſuit,
In Studiis nullum.

Quid

"bestowed on his friend and neighbour,
"Mr. *Samuel* DALE, author of the *Phar-*
"*macologia*, to whom they were delivered
"about a week before his death."

Mr. RAY's posthumous papers were en-

Quid plura?
Hisce omnibus
Pietatem minimè fucatam adjunxit,
Ecclesiæ Anglicanæ
(Id quod supremo Habitu confirmavit)
Totus et ex Animo addictus.
Sic bene latuit, bene vixit Vir beatus,
Quem præsens Ætas colit, Postera mirabitur.

This monument beginning to want repair by standing exposed in the church-yard, was removed and set up in the chancel of the church; and to the epitaph is added, on the table of the east side, what follows:

Hoc Cenotaphium
Olim in Cœmeterio sub Dio positum,
Inclementis Cœli Injuriis obliteratum,
Et tantum non collapsum,
Refecit et sub Tectum transposuit
J. LEGGE, M. D.
xvi kal. Aprilis, A. D. 1737.

On the west side,

J. RAY, { Nat. 29. Nov. 1628.
Ob. 17. Jan. 1705-6. }

 trusted

truſted by his widow to the care of Dr. DERHAM; who, after publiſhing the "HISTORIA INSECTORUM," ſelected a number of his letters, and printed them, in 1718, under the title of "PHILOSOPHICAL LETTERS between the learned Mr. RAY and "ſeveral of his Correſpondents, natives "and foreigners." 8°. pp. 367.

This collection contains 218 letters; of which, ſixty-eight were written by Mr. RAY himſelf. Among his correſpondents, the moſt frequent were Dr. LISTER, Sir *Philip* SKIPPON, Dr. *Tancred* ROBINSON, Sir *Hans* SLOANE, Mr. LLWYD, Mr. JESSOP, Mr. JOHNSON, and Mr. OLDENBURGH. The firſt of Mr. RAY's letters bears date in 1667, the laſt in 1705.

The correſpondence of learned and ſcientific men, ſeldom fails to be a welcome preſent to thoſe of ſimilar literature and purſuits; for, beſides the perſonal intereſt we take in their concerns, they commonly delineate, in the moſt faithful colours, the characters of the writers, frequently aſcertain diſcoveries, and enable their ſucceſſors to trace the progreſs of knowledge in a

more

more intereſting manner than by hiſtorical detail.

As the general ſubject of theſe letters is natural hiſtory, ſo botany bears a prevailing portion. Beſides numberleſs critical obſervations that occur on particular ſpecies, we meet with a long catalogue of the rare plants of the north of *England*, by Mr. LAWSON; Dr. PLUKENET's Obſervations on the firſt edition of the "*Synopſis*;" thoſe of Dr. PRESTON on various *Britiſh* Plants; a paper of *Thomas* WILLISEL's ſpecifying the different kinds of trees, on which, in his travels, he had ſeen the *Miſſeltoe* growing; and a liſt of ſuch exotics as were thought rare at that time in the *Chelſea* Garden, and at *Fulham*.

There is, moreover, among theſe letters, an intereſting paper, written by Mr. RAY himſelf, in anſwer to the queſtion, "What "number of plants there are in the world?" in which he diſcuſſes the difficulty, or impoſſibility, of gaining ſatisfaction on this point, ariſing from the want of ſufficient bounds between ſpecies and variety. He communicated to the Royal Society ſome remarks on this head, which were printed by Dr. BIRCH, in the third volume of the "Hiſtory of the Royal Society."

Dr. DERHAM meditated writing the life of Mr. RAY; but he appears not to have fully executed his plan. His papers, however, were publiſhed by Mr. SCOTT, in 1760, under the title of "Select Remains "of the learned *John* RAY." 8°. pp. 336. To theſe are annexed three of the *Itineraries*, which conſtitute the greater part of the book. They are evidently ſhort notes only, never intended for the public eye. Some of Mr. RAY's devotional pieces accompany this collection; and three letters to Dr. DERHAM; with a *Latin* letter of advice and inſtructions to his pupils, the Mr. WILLUGHBYS.

There is ſaid to be ſtill extant a manuſcript of Mr. RAY's, under the title of "*Catalogus Plantarum domeſticarum quæ aluntur Catabrigiæ in hortis academicorum et oppidanorum.*" In this, he chiefly makes uſe of the *ſynonyma* of the two BAUHINES, and of GERARD and PARKINSON.

Mr. RAY had the ſingular happineſs of devoting

devoting fifty years of his life to the cultivation of the ſciences he loved. Incited by the moſt ardent genius, which overcame innumerable difficulties and diſcouragements, his labours were, in the end, crowned with a ſucceſs, before almoſt unequalled. He totally reformed the ſtudies of botany and zoology; he raiſed them to the dignity of a ſcience, and placed them in an advantageous point of view; and, by his own inveſtigations, added more real improvement to them in *England*, than any of his predeceſſors.

He invented and defined many terms, expreſſive of ideas before unknown to the naturaliſts of *England*; and introduced many others, from writers of the beſt note. As he wrote *Latin* in great purity, and with great facility, he gave his ſubjects all the embelliſhments that learning could beſtow; and his extenſive erudition, and knowledge of philoſophy at large, enabled him to add many collateral ornaments, and uſeful obſervations, with an aptitude and judgment that has been much applauded.

The extent of his improvements in ſcience procured him the admiration of his contemporaries, and have juſtly tranſmitted his name to poſterity, among thoſe who have done honour to their age and country. Even learned foreigners have been eloquent in his praiſe. *French* writers have ſtiled him the "*Engliſh* TOURNEFORT;" an eulogy that ſufficiently evinced the high opinion they had of his merit. And the late eminent HALLER not only attributes to RAY the merit of improving and elevating botanical knowledge, but from his life dates a new æra in the records of the ſcience.

But Mr. RAY's enquiries were not limited to natural knowledge. His Foreign Travels and his Itineraries prove, that antiquities, polity, government, and legiſlation, attracted a ſhare of his regard; as his philological books are evidences of his attention to language, and of his deſire to improve and illuſtrate his native tongue.

To all theſe endowments he joined an unremitting induſtry and perſeverance in the proſecution of his ſtudies; and, what marks a fortitude of mind as uncommon as it is enviable, his aſſiduity ſeemed to ſtrengthen with

with his age, and to bid a defiance to the encroachments of infirmity, and the prospect of dissolution. I call to witness the magnitude of the attempt, and successful issue of his exertions, in writing the supplemental volume to his " History of Plants," and in beginning the " *Historia Insectorum* " at so late a period of his life.

His singular modesty, affability, and communicative disposition, secured to him the esteem of all who knew him; and his eminent talents as a naturalist and a philosopher procured him many patrons and friends, and preserved him from that obscurity, which would otherwise probably have been his lot: for, notwithstanding his learning and probity, as his principles did not accord with those of the times, they were adverse to his fortune, and he gained no emoluments in the church. He had relinquished his fellowship at the commencement of the *Bartholomew* act, not, as some imagined, from his having taken the *Solemn League and Covenant* (for that he never did, and often declared, that he ever thought it an unlawful oath), but because he could not declare, agreeably to the terms of the act, that the oath was not binding on those who had taken it. Hence too, his constant refusal of preferment afterwards, occasioned him to be ranked, by many, among the nonconformists, although he lived and died in the communion of the church of *England*. He had seen, with deep regret, the disorders of the commonwealth and the usurpation, and afterwards, not less, the threatening aspect of the reign of *James* II.

His strong attachment to the principles of civil and religious liberty, is manifested by his animated stile, in the preface to his "*Synopsis*;" where he expresses, in glowing terms, his joy and gratitude, for having lived to see those blessings established by the Revolution.

The character of Mr. RAY cannot be contemplated by those who have a true relish for the studies of nature, without a high sentiment of respect and gratitude; nor by those who consider the exemplariness of his life as a man, and his qualifications as a divine, without veneration.

There are two engraved portraits of Mr.

RAY

RAY prefixed to his works, both from a painting by *Faithorne*; one by *W. Elder*, before his "SYLLOGE," in 1693, which ſeems to have been copied for the "METHODUS EMENDATA," in 1703; and the other by *Vertue*, in 1713, prefixed to the "Phyſico-theological Diſcourſes." In both theſe, he is repreſented, as Mr. AMES deſcribes it, in "an oval frame, with hair, "whiſkers, band, and canonical habit." Theſe engravings repreſent Mr. RAY in the latter ſtage of his life *.

* In dedicating plants to the worthies of botanical ſcience, the name of RAY challenged a dignified place; and the liberal-minded foreigner, whoſe name has before occurred on theſe occaſions, forgot not ſo juſt a tribute. PLUMIER called a new plant of the *dioecious* claſs, which bears the habit of *bryony*, and is nearly allied to the *yams*, which he firſt diſcovered in the iſle of *Domingo*, by the name of JAN-RAJA, in honour of our illuſtrious countryman. LINNÆUS, who had comparatively few opportunities of correcting PLUMIER, eſtabliſhed the *genus*, but more aptly changed it to RAJANIA, and enumerates three ſpecies. He could not adopt the ſtill more analogous term of RAIA, ſince it had long been preoccupied in the animal kingdom; and it had been juſtly conſtituted an axiom, by the *Fundamenta Botanica*, N° 230, not to form, in the vegetable kingdom, any generical terms, ſynonymous to ſuch as were employed in zoology or mineralogy.

CHAP.

CHAP. 21.

Poetical botaniſts—Cowley—*Account of his poems on plants*—*Not deeply verſed in the botany of his time*—*Intimate knowledge of natural hiſtory neceſſary to accompliſh* "*the poet of nature.*"

COWLEY.

IN all times, from VIRGIL and ÆMILIUS MACER of the Auguſtan age, from the ſpurious MACER, and STRABUS the monk of *St. Gall*, in the twelfth century, to modern times, the beauties of flowers, and the virtues of plants, have been celebrated in verſe. *Marcus* NÆVIANUS, firſt a phyſician, and then a prieſt, of *Flanders*, ſung the qualities of plants in his "*Poemation*" of 1563; and THUANUS, the great hiſtorian, amuſed himſelf with praiſing the *violet* and the *lily* in metre. In our own country, in 1723, *George* KNOWLES deſcribed 400 plants of the *Materia Medica*, in *Latin* verſe, and didactically applied them to their uſes in medicine.

But to proceed: That *England* and *France*, in

in the ſame age, might not want their botanical laureats, COWLEY in the one, and RAPIN in the other, aroſe to celebrate this theme.

COWLEY, after having found reaſons for ſtudying phyſic, " conſidering botany," as we are told by his late eminent biographer, " as neceſſary to a phyſician, retired into " *Kent*, to gather plants."

Here, he wrote, before the Reſtoration, his " Two firſt Books on Plants;" although they were not publiſhed till the year 1662. The remaining four were added in the edition of 1668; and the whole were republiſhed, with other poems, in 1678. 8. pp. 343.

In the *firſt* book, he celebrates the powers of various medicinal herbs, more eſpecially of thoſe which gave ampler ſcope to his muſe, from antient renown of their virtue, and were yet in frequent uſe, and high eſteem. Such were betony, wormwood, water lily, miſſeltoe, and various others.

In the *ſecond*, he invokes the goddeſſes *Luna*, *Lucina*, *Jana*, and *Mena*; and ſings the praiſes of ſimples appropriated to the diſeaſes of the ſex: in which, both antient ſuperſtition, and modern belief, ſupplied his muſe with exuberant ſources of gratification.

In the *third*, *Flora* calls forth all his powers, in the narciſſus, the anemone, the violet, and the tulip, with a variety of other ornaments of the parterre, from the *coronary* tribe.

In the *fourth*, a more numerous ſet of the ſubordinate embelliſhments of the garden are recorded, in various meaſure; among which, the attributes of the moly, the lily, poppy, ſunflower, ſaffron, and amaranth, attract his muſe with more than ordinary attention.

In the *fifth*, he celebrates, in heroic meaſure, the gifts of *Pomona*, from the native products of *England*, to the date of the eaſt, and the *tuna* of the weſt; terminating his poem with near two hundred lines on *Columbus*, on the *Spaniards*, on the new continent, and in expreſſing his hopes that, to the devaſtations of conqueſt, will ſoon ſucceed peace, religion, arts, and ſcience.

In the *laſt*, he diſplays the ſylvan ſcene, from

from the oak of *Boscobel*, to the lowly juniper; and, having constituted his druidical monarch the sovereign of the forest, he makes him the oracle for a train of reflections, on the usurpation; the exile of *Charles* the Second, his restoration; and the *Dutch* war.

His poems are accompanied by notes, illustrating the etymology, the names, synonyms, descriptions, faculties, and uses of the plants, confirmed by authorities drawn from classical, botanical, and medical writers. Of these, he professes in his preface, that PLINY among the antients, and FERNELIUS among the moderns, have been his chief resources. Of botanical authors, GERARD and PARKINSON are sparingly mentioned, and they are the principal of that class.

Great eminence in science is seldom attainable, unless its foundation be laid in a devotedness of mind to its object, in the early scene of life. COWLEY did not enter on the study of physic, till the middle age of man; and then, as is probable, not with interested views towards practice. Hence it it may fairly be presumed, that he satisfied himself with moderate acquisition. What was true of the whole, may by fair analogy be applicable to a particular branch of it. He had doubtless that portion of knowledge in the *materia medica* of plants, which may be considered as adequate to the usual demand.

But, that COWLEY, in his retirement, should obtain an extensive and critical knowledge of botany, as it stood as a science, even in his day, could not be expected. His fervid genius could scarcely stoop to that patient investigation of nature, by which alone it could be acquired. Neither do the text, nor the notes, manifest sufficient proof of his intimate acquaintance with those authors of true fame, among the moderns, through whose assistance the want of that information might, in some measure, have been supplied.

Nevertheless, as, in the language of Dr. JOHNSON, "Botany, in the mind of COWLEY, turned into poetry," to those who are alike enamoured with the charms of both, the poems of COWLEY must yield delight;

delight; ſince his fertile imagination has adorned his ſubject with all the beautiful alluſions that antient poets and mythologiſts could ſupply; and even the fancies of the modern *Signatores*, of BAPTISTA PORTA, CROLLIUS, and their diſciples, who ſaw the virtues of plants in the phyſiognomy, or agreement in colour or external forms, with the parts of the human body, aſſiſted to embelliſh his verſe. Nor did he fail, by theſe elegant productions, to honour his ſubject, his name, and his country.

I cloſe theſe obſervations by remarking, that poetry, as it ever hath, ſo it ever muſt derive from nature ſome of its moſt pleaſing ſcenes of entertainment. In the vegetable world, the moſt expanded imagination of poetic genius will, even without the aid of fiction, ſo emphatically ſtiled the ſoul of poetry, find a field ſufficiently ample for the diſplay of the brighteſt talents. THOMSON witneſſes this truth, while in him we lament the want of that botanical knowledge, without which, the poet muſt ever be deprived of numberleſs ſources of the

moſt

moſt beautiful imagery, and ſuch as would add peculiar grace, and the moſt inſtructive power to his muſe.

And, although the talent of the poet hath not often been united to that of the really ſcientific botaniſt, there are not wanting inſtances of this union. I might mention, ſince the diſcovery of the ſexes of plants, the ode, dedicated to CAMERARIUS, and printed in his "*Epiſtola de Sexu Plantarum*;" of which, a tranſlation by Dr. MARTYN, when a young man, may be ſeen in BLAIR's "Botanick Eſſays." Profeſſor *Van* ROYEN, in 1732, publiſhed an elegant poem "*De Plantarum Amoribus, et Connubiis.*" And CUNO, an ingenious merchant of *Amſterdam*, in a volume of 256 pages, deſcribed, in 1750, the plants of his own garden in verſe; for which he received the laurel from LINNÆUS, by a new genus inſcribed to his name.

Whilſt I am now writing, I have the pleaſure of congratulating all thoſe, whoſe love of poetry is aided by a taſte for botanical ſcience, on a moſt elegant production in

our

our own country. The beautiful diſplay of the principles of the *Linnæan* ſyſtem in the "Botanic Garden," under the delicate analogy of the "Loves of the Plants," in which the didactic deſign of the author, is ſo happily embelliſhed by *Ovidian* imagery, as to have given that energy and ornament to the ſubject, which has been hitherto wanting to all ſimilar productions in the *Engliſh* language.

CHAP. 22.

Merret, *brief anecdotes of—His* Pinax Rerum Naturalium, *intended to ſupply the deficiencies of* How's Phytologia —*Aſſiſted by* Williſel: Goodyer's *manuſcripts* – Merret's *other writings—His papers in the* Philoſophical Tranſactions.

MERRET.

" CHRISTOPHER, the ſon of *Chriſ-*
" *topher* MERRET, was born at
" *Winchcombe*, in *Glouceſterſhire*, Feb. 16,
" 1614. He became a ſtudent in *Glou-*
" *ceſter* Hall, in the beginning of the year
" 1631; two years after which time, he
" tranſlated himſelf to *Oriel* College, and
" took the degree of B. A. in 1634. Af-
" terwards, retiring again to *Glouceſter*
" Hall, he applied to the ſtudy of phyſic,
" and was created doctor in that faculty in
" 1642. About this time he ſettled in
" *London*, and came into conſiderable prac-
" tice, was a fellow of the College of
" Phyſicians, and of the Royal Society.

" He

" He died at his houſe, near the chapel in
" *Hatton Garden*, in *Holborne*, near *London*,
" Aug. 19, 1695; and was buried twelve
" feet deep in the church of *St. Andrew's*,
" *Holborne*." Thus far Mr. *Wood*.

The publication which entitles Dr. *Merret* to a place in theſe anecdotes, is, his " PINAX RERUM NATURALIUM BRITANNICARUM, *continens* VEGETABILIA, *Animalia, et Foſſilia, in hac Inſula reperta*." 8°. 1667. pp. 223.

This is not noticed in the title as a ſecond edition, although there is one recorded by authors, with the date of 1665. However, I ſuſpect it to be a miſtake, as no ſuch edition is quoted by RAY. He dates his book from the College of Phyſicians, and is mentioned by MORISON under the title of " *Muſei Herbiani Cuſtos*."

Dr. MERRET informs us, that he undertook this work at the requeſt of a bookſeller, to ſupply the deficiencies of How's " *Phytologia*," after that work was out of print; and that it was intended to have been done jointly with Dr. DALE, whoſe death, ſoon after the deſign was formed, threw the whole into his own hands. He ſays, he had purchaſed 800 figures, which JOHNSON had cauſed to be engraved, with which the work was to have been embelliſhed. Why they did not appear, no cauſe is aſſigned; nor do I find any further notices of them. Dr. MERRET, though unqueſtionably a man of learning, taſte, and conſiderable information in natural hiſtory, ſeems to have engaged in it too late in life, to admit of his making that proficiency, which the deſign required. Add to this, that being fixed in *London*, and cloſely engaged in the practice of his profeſſion, he was rendered incapable of inveſtigating plants, in the diſtant parts of the kingdom. He however engaged *Thomas* WILLISEL to travel for him; and he tells us, that WILLISEL was employed by him for five ſucceſſive ſummers. His ſon, *Chriſtopher* MERRET, alſo made excurſions for the ſame purpoſe; and Mr. *Yauldon* GOODYER furniſhed him with the manuſcripts of his grandfather. By theſe aſſiſtances Dr. MERRET procured a large number of *Engliſh* plants, and a knowledge of the *Loci Natales*.

Never-

Neverthelefs, he was not poffeffed of that critical and intimate acquaintance with the fubject, which might have enabled him to diftinguifh, with fufficient accuracy, the fpecies from varieties. He ranges the plants alphabetically, according to the *Latin* names, and has given few fynonyms, except thofe of GERARD and PARKINSON; to which, after the example of the writers of the "*Hortus Oxonienfis*," he has very commendably annexed the page. He gives the general places of growth, and fpecifies the particular fpots, where the rare plants are found.

At the end of the Catalogue, is fubjoined, a rude difpofition of vegetables into claffes, fomewhat like that of *John* BAUHINE. This he hoped to have improved, againft the time of a fecond edition, which, probably, Mr. RAY's publications fuperfeded. Then follows a brief *Synopfis Etymologica*, and a ufeful lift of the plants as they flower in each month, pointing out the duration of the time. Dr. MERRET has, in this *Pinax*, introduced many plants as new, which, on fubfequent examination, proved to be only varieties; a number of exotics, evidently the accidental offspring of gardens, and many that could never be met with by ſucceeding botaniſts, in the places ſpecified by him. He enumerates upwards of 1400 ſpecies of *Engliſh* plants; whilſt the accurate Mr. RAY, only three years afterwards, confines the number to 1050. Nevertheleſs, ſeveral *Britiſh* plants make their firſt appearance in this *Pinax*; and Dr. MERRET would probably have ſecured his title to ſome others, if he had not totally omitted to give deſcriptions of thoſe which he introduces as new.

The zoological part of this *Pinax* is extremely ſuperficial; conſiſting merely of the *Latin* and *Engliſh* name, with a reference to ALDROVANDUS, GESNER, JOHNSTON, and MOUFFETT. The mineralogy is not leſs brief, and imperfect.

Before the publication of this work, Dr. MERRET had printed "A Collection of "Acts of Parliament, Charters, Trials at "Law, and Judges Opinions, concerning "thoſe Grants to the College of Phyſi- "cians." 4°. 1660. This became the baſis,

ſis, as Mr. *Wood* ſays, of Dr. GOODALL's book, printed in 1684.

In 1669, he wrote "A ſhort View of "the Frauds and Abuſes committed by "Apothecaries, in relation to Patients, and "Phyſicians." 4°. This treatiſe engaged him in a controverſy with the famous *Henry* STUBBE. It may be preſumed, that all diſcuſſions of this kind, howſoever well meant, can have but little effect in reforming the abuſes hinted at, while the cuſtomary and legal conſtitution, and polity of phyſic, remain in the preſent ſtate in *Great Britain*.

In 1662, he tranſlated into *Engliſh*, "The "Art of Glaſs; how to colour Glaſs, Ena-"mels, Lakes, &c." 8°. written by *Ant.* NERI, accompanied with an account of the Glaſs Droſs. And, in 1686, the ſame work was publiſhed in *Latin*, with Dr. MERRET's "Obſervations and Notes," equal in extent to the work itſelf. *Amſt.* 12°.

Mr. *Wood* informs us, that he alſo printed, in one ſheet, 4°. "The Character of a "compleat Phyſician or Naturaliſt."

Dr. MERRET was among the earlieſt

members of the Royal Society, after its incorporation; and contributed ſeveral papers, which were printed in the "*Philoſophical* "*Tranſactions*."

He made experiments on vegetation, in the year 1664; by which he found, that ſquare ſections of the bark, from aſh, and maple, whether ſeparated on three ſides only, or wholly, would firmly unite, if tightly ſecured by plaiſter and packthread.

Experiments on the loſs of weight, which a plant of the *Aloe Americana*, with eleven leaves, ſuffered by hanging up in the kitchen for five years. In the firſt year it loſt near two ounces and an half; the ſecond upwards of three ounces; decreaſing afterwards nearly in the ſame proportion. It loſt two of the larger leaves every year, and put forth two new ones every ſpring; from which circumſtance, the Doctor inferred a circulation of the juice.

Experiments on cherry-trees, that, having withered fruit, occaſioned by the ſun being admitted too ſuddenly upon them in March, recovered, by daily watering the roots.

Obſervations

Obſervations on the *London* granaries. Theſe four papers were all printed in N° 25, in the ſecond volume of the "*Tranſactions.*"

In N° 138, an account of the tin-mines in *Cornwall*, mundic, ſpar, and *Corniſh* cryſtals.

In N° 142, an account of the art of refining, in the ſeveral methods, by parting, by the teſt, the almond furnace, and by mercury.

In N° 223, ſome curious obſervations on the fens of *Lincolnſhire*; on the animal and vegetable produce: a deſcription of *Boſton* church, the incroachments of the ſea, and other particulars, which muſt have rendered this paper a very intereſting morſel of natural hiſtory. He gives a liſt of ſeveral of the more rare plants growing in the fens.

In N° 224, a table of the waſhes called *Foſdyke* and *Croſskeys*, in *Lincolnſhire*, ſpecifying the times of high water, and ſafe paſſage over the ſands.

CHAP.

CHAP. 23.

Moriſon—*Account of his life—His* Hortus Bleſenſis; *in which are contained the rudiments of his ſyſtem, and the animadverſions on the* Bauhines—*Publiſhes* Boccone's Plantæ Siculæ—*His* Diſtributio Plantarum Umbelliferarum—*His great work, the* Hiſtoria Plantarum Oxonienſis—*Outlines of his method.*

Jacob Bobart, *the continuator of* Moriſon's Hiſtory—*Brief anecdotes of.*

MORISON.

ROBERT MORISON was born at *Aberdeen*, in 1620; was educated in the ſame univerſity; and, in 1638, took the degree of doctor in philoſophy, equivalent to that of M. A. He firſt applied to mathematics, and was deſigned by his parents for the theological line; but his taſte for botany and phyſic ſuperſeded their intentions. His attachment to the royal cauſe, led him into the army; and he received a dangerous wound in the head, in the battle at *Brigg*, near

near *Aberdeen*. Upon his recovery, he went to *Paris*, the aſylum of his countrymen. Here he was firſt employed as a tutor to the ſon of a counſellor, *Bizet*; and, in the mean time, aſſiduouſly applied to the ſtudy of anatomy, botany, and zoology. In 1648, he took the doctor's degree in phyſic at *Angers*. He became ſo much diſtinguiſhed by his ſkill in botany, that at the recommendation of M. ROBINS, the king's botaniſt, he was taken into the patronage of the Duke of *Orleans*, uncle to *Lewis* XIV. and appointed intendant of his fine garden at *Blois*, with a handſome ſalary. This eſtabliſhment took place in 1650, and he held it until the death of the Duke, in 1660. Here, we are told, MORISON laid open to the Duke his method of botany; and was liberally encouraged by him to proſecute it. The Duke alſo ſent him into various provinces of *France*, to ſearch for new plants. He travelled into *Burgundy*, the *Lyonnois*, and *Languedoc*; and into *Britanny*, the coaſts and iſles of which he carefully inveſtigated; and, by theſe journies, enriched the garden with many rare, and ſome new plants.

It

It was in this ſituation that he became known to *Charles* II. who, in 1660, on the death of his uncle the Duke, invited MORISON into *England*; and, although ſolicited by the treaſurer *Fouquet*, on the moſt honourable and ample conditions, to remain in *France*, the love of his country overcame all temptations, and he returned to *England*. *Charles* II. gave him the title of king's phyſician, and royal profeſſor of botany, with an appointment of 200l. a year, and a houſe, as ſuperintendant of the royal gardens. He was elected fellow of the Royal College of Phyſicians, and acquired much fame for his knowledge of botany. In this ſituation he remained till the year 1669, when, having made an acquaintance with Mr. *Obadiah* WALKER, of Univerſity College, with the Dean of Chriſt Church, and other leading men of the univerſity, he was, by their intereſt, elected botanic profeſſor at Oxford, Dec. 16, 1669, and incorporated doctor of phyſic the day following. He read his firſt lecture in the phyſic ſchool in September 1670, and then removed to the phyſic garden, where he lectured

lectured three times a week, to a conſiderable audience. In this occupation, and in conducting his great work, the "*Hiſtoria Plantarum Oxonienſis*," he laboured to the time of his death, which was thought to have been occaſioned by a bruiſe, received by the pole of a coach, in croſſing the ſtreet, Nov. 9, 1683. He died at his houſe in *Green-ſtreet, Leiceſter Fields,* the next day, and was buried in the church of *St. Martin's in the Fields, Weſtminſter.*

SEGUIER ſeems to have placed improperly among MORISON's works the firſt edition of the "*Hortus Bleſenſis*," which he gives as publiſhed in the year 1635, when MORISON muſt have been only fifteen years of age. This may have been a typographical error; but the book, in fact, was the work of *Abel* BRUYNER, phyſician to the Duke of *Orleans*, and was not publiſhed till 1653. MORISON's firſt publication was a ſecond edition of this catalogue, under the following title: "HORTUS REGIUS BLESENSIS *auctus: acceſſit Index Plantarum in Horto contentarum nemini Scriptarum et Obſerva-*

Obſervationes generaliores, ſeu Præludiorum pars prior." Lond. 1669. 12°.

The "HORTUS BLESENSIS" raiſed the author's character, and contributed, as the writer of his life obſerves, to recommend him to the ſtation he afterwards held at *Oxford*. It contains the rudiments of his method of claſſification. He profeſſes to give a liſt of 260 new plants; but many of them proved to be only varieties, and others, ſuch as were well known before. There were, nevertheleſs, ſome new and rare plants, of exotic, as well as indigenous origin; the latter, ſuch as he had himſelf firſt diſcovered in *France*.

In this work is alſo given his "HALLUCINATIONES *in* CASPARI BAUHINI *Pinacem, tam in digerendis quam denominandis Plantis; et his Animadverſiones, in tres Tomos, Hiſtoriæ Plantarum* JOHANNIS BAUHINI;" a work which *Haller* calls "*Invidioſum Opus;*" and which, while it proves both the accuracy and diligence of the author, muſt be confeſſed to be unbecomingly ſevere on theſe two illuſtrious writer ; who,

as

as they did not profeſs to write a ſyſtem, are here too rigidly tried by rules, not invented when they wrote, and of conſequence the validity of which they could not have acknowledged.

In a dialogue at the end of the "*Hortus Bleſenſis,*" MORISON teaches, that the genera of plants ſhould be eſtabliſhed on characters drawn from the fruit, and not on any ſenſible qualities, or ſuppoſed medicinal virtue. He alſo learnedly defends the doctrine, that all vegetables ariſe from ſeed; a propoſition not univerſally allowed; the doctrine of equivocal, or ſpontaneous generation, having, at that time, many advocates among the learned.

Dr. MORISON, during his reſidence in *France*, in his occaſional journies to *Paris*, about the year 1658, became familiar in the family of Lord HATTON, then reſident at *St. Germains*, and whoſe ſecond ſon *Charles* was much attached to natural hiſtory, and became a voluntary and zealous diſciple of our author. Sixteen years afterwards, Mr. *Charles* HATTON ſent over, at the author's requeſt, a treatiſe, with the plates already engraved,

graved, written by *Paul* BOCCONE, on plants, discovered by him in the southern parts of *Europe*, principally in *Sicily*, of which some were rare, and some new. BOCCONE was originally of *Savona*, in the *Genoese* district; and was born in 1633. He became a *Cistertian* monk of *Palermo*, and was a man of singular and various erudition in natural history. He visited *Corsica* and *Malta*; travelled into *England*, *Holland*, and *Germany*; and was for some time botanist to the Duke of *Tuscany*. He was the author of several very curious works; and died in 1704. He wrote on fossils; but his botanical writings have greater originality, and were of high value. MORISON, after having caused the seven last plates to be re-engraved, published the work alluded to above, under the following title:

"ICONES *et* DESCRIPTIONES RARIORUM PLANTARUM *Melitæ, Galliæ, et Italiæ. Auctore Paulo* BOCCONE, *panormitano siculo, serenissimi magni Etruriæ Ducis olim Botanico.*" Oxon. 1674. 4°. pp. 96. t. 52. fig. 119.

MORISON prefixed to this work a dedication

cation to Mr. HATTON, in which he defends, not only the doctrine in general, that all plants ſpring from ſeed, but particularly, againſt DIOSCORIDES, and ſome of the reſtorers of ſcience, among whom were CÆSALPINUS, that all the ferns are furniſhed with flowers and ſeed.

The plants deſcribed and figured in this book, are, moſt of them, ſuch as had not been noticed by foregoing authors. A few of theſe are common to *Britain*. The figures are ſmall, and neither well delineated, nor well engraven: but the work had its uſe, as containing ſome plants of Southern *Europe*, not to be met with in any other author; and on this account derives ſome value, to thoſe who are curious in purſuing the hiſtory of plants in the ſexual ſyſtem, as being quoted by LINNÆUS.

As a ſpecimen of his great work, meditated under the name of "*Hiſtoria Plantarum Univerſalis Oxonienſis*," MORISON next publiſhed, "PLANTARUM UMBELLIFERARUM DISTRIBUTIO NOVA, *per tabulas cognationis et affinitatis, ex libro Naturæ obſervata et detecta*." Oxon. 1672. fol.

pp. 91. t. 12. The umbelliferous tribe is here divided into nine orders, the genera of which are diſtinguiſhed by the figure of the ſeed, aſſiſted, in ſome of the ſubdiviſions, by the form of the leaf. They are illuſtrated by figures of 150 different ſeeds.

The author has ſubjoined what he names "Umbelliferous Plants, improperly ſo called." Such are *Valeriana, Thalictrum, Filipendula, Valeriana græca, Pimpinella Sanguiſorba*; all which are very different, both in character and habit, except the *Valerian*, from the natural claſs of which he treats.

This ſpecimen excited the attention of the learned, augmented MORISON's patronage, both abroad and at home; and encouraged him to proſecute with vigour his great work, of which the firſt volume came out under the following title: "PLANTARUM HISTORIÆ UNIVERSALIS OXONIENSIS, *Pars ſecunda; ſeu Herbarum Diſtributio nova, per tabulas cognationis et affinitatis, ex libro Naturæ obſervata et detecta.*" Fol. 1680. pp. 617. The firſt part of the Hiſtory, on Trees and Shrubs, was never printed. Some have doubted, whether it was

was ever written; but SCHELHAMMER* tells us, that he ſaw the whole work perfect in the hands of the author. MORISON himſelf aſſigns, as a reaſon for publiſhing the Herbaceous Diviſion firſt, the greater magnitude of the undertaking, ariſing from the vaſt number, and conſequent difficulty of finding proper diſtinctions and characters; and becauſe he was unwilling to leave the moſt difficult and abſtruſe part of his work behind him unfiniſhed, as happened to DELECHAMP, and *John* BAUHINE. Unhappily, however, MORISON's untimely death ſubjected his work to the ſame lot, and did not allow him to finiſh more than nine, out of the fifteen claſſes of his own ſyſtem.

He divides all herbaceous plants into fifteen claſſes, under the following titles:

1. *Scandentes.*
2. *Leguminoſæ.*
3. *Siliquoſæ.*
4. *Tricapſulares Hexapetalæ.*
5. *Tricapſulares, aliæ.*
6. *Corymbiferæ.*
7. *Pappoſæ Lacteſcentes.*
8. *Culmiferæ.*
9. *Umbelliferæ.*
10. *Tricoccæ Purgatrices.*

* In additamentis ad CONRINGIUM.

11. *Galeatæ, et Verticillatæ.*
12. *Multisiliquæ, et Multicapsulares.*
13. *Bacciferæ.*
14. *Capillares.*
15. *Anomalæ.*

From an inspection of this table, it appears, that his method is not uniformly founded on the fruit; in fact, much less so than that of CÆSALPINUS; but on the fruit and the habit conjointly; since the *Corymbiferæ*, *Umbelliferæ*, and *Galeatæ*, with the *Verticillatæ*, arise from the disposition of the flower; the *Scandentes*, *Culmiferæ*, and *Capillares*, from the habit: the seventh class from the qualities partly, and partly from the seed. Hence we see, that only half the classes are founded on the fruit; the fifteenth being truly an heteroclite assemblage. His method would have approached much nearer to perfection, on his own principles, had he enlarged the number of his classes; since, in several instances, they embrace natural orders, much too distinct to be ranged together. The orders, or subdivisions of the classes, are, in some instances, grounded on differences in the

seed-

seed-vessel; in others, on the root, habit, and frequently on less scientific discriminations. In the conduct of the work itself, MORISON makes a separate chapter for each genus. He begins by referring to the antients under each plant; frequently subjoining the etymology. The generical characters, if indeed they can be so called, are very vague; and though taken from the parts of fructification, are, too often, assisted by distinctions from the root, leaves, and mode of growth. After the generical note, follows a synoptical table of the species, referring to the plates. The descriptions are sometimes borrowed from *John* BAUHINE and others. To most of the plants, he affixes new specific characters, and subjoins the *synonyma* of several authors. He introduces, at the end of the chapters, the animadversions on the BAUHINES, and an account of the virtues and uses of the plants.

The five first classes only, were published by the author, who left the four succeeding ones finished. These, with the remaining classes, were finished and published, after an interval of nineteen years, by *Jacob* Bo-

BART. MORISON had the advantage of powerful patronage. He was liberally encouraged by the univerſity, and enabled to embelliſh his work with a numerous ſet of tables, on which are engraven about 3384 plants. The figures are chiefly copied from other authors. The new figures occur principally in the latter part of this work, and are therefore to be attributed to the care of BOBART. The ſix tables of *Moſſes*, *Fuci*, *Corallines*, and *Corals*, at the end, are, except the few wooden cuts of GERARD, the firſt of the kind graved in *England*, and have great merit as the productions of that time. All thoſe of COLUMNA and CORNUTUS are copied in this work. Thoſe engraved by *Burghers* excel the reſt; and the figures of the graſſes and moſſes are incomparably beyond any other that are to be met with, on the ſame ſcale; the habit being admirably well expreſſed. The republication of theſe tables, with references to LINNÆUS's writings, would, even at this period, be a benefit to the ſcience.

The third part, or, more properly, the ſecond volume of the "Oxford Hiſtory of "Plants,"

" Plants," was publiſhed by *Jacob* BOBART, in fol. 1699. pp. 655. A life of MORISON is prefixed to this volume, and an engraving of him done by *White*, with Dr. PITCAIRN's Tetraſtic underneath. In the preface, which is ſigned *Jacob* BOBART, the reader is preſented with a general view of the writers on botany, from THEOPHRASTUS, to the time of MORISON; enumerating throughout the ſeveral nations of *Europe*, in a chronological order, the moſt learned authors on the ſubject. The writer then informs us of the patronage and encouragement which MORISON received from the univerſity, to undertake this work; and, after lamenting the untimely death of the author, and expreſſing his grateful ſenſe of the honour he received in being appointed to continue the undertaking, he lays before the reader the aſſiſtances he received in the proſecution of it. An interval of near twenty years had given BOBART an opportunity of inſerting a great number of plants unknown to MORISON, from the works of RAY, HERMAN, PLUKENET, the "*Hortus Malabaricus*," and other works.

 With

With reſpect to *Engliſh* botany, great communications had been made by SLOANE, PETIVER, DOODY, SHERARD, and others. By theſe means, this volume contains nearly double the number of the former; but the latter part of it proves, too evidently, that it did not receive the finiſhing hand of the original author; ſince it appears in a very abridged form, compared with what MORISON * himſelf had done.

BOBART.

Jacob BOBART, the continuator of MORISON'S Hiſtory, was the ſon of *Jacob*, the firſt ſuperintendant of the Garden, upon its foundation in 1632. Both the father and ſon filled their ſtation with great credit to themſelves, and no leſs emolument to the Garden. The elder is ſaid to have been the author of the firſt edition of the "*Hortus Oxonienſis*," 1648; and his name is joined in the ſecond edition, 1658, as an aſſociate in the work, with Dr. STEPHENS and Mr.

* The name of MORISON is perpetuated by PLUMIER, in the application of it to a Weſt Indian tree of the *monadelphous* claſs, hitherto deſcribed only by himſelf and JACQUIN.

BROWNE.

BROWNE. Mr. GRANGER relates a humorous circumſtance in his manners; that "on "rejoycing days, he uſed to have his beard "tagged with ſilver." He died in 1679, at the age of eighty-one; and left, beſides *Jacob*, another ſon, named *Tillemant*, who was alſo employed in the Phyſic Garden.

I cannot aſcertain the time of BOBART's death; but from the ſtory related of him by Dr. *Grey*, in his edition of "Hudibras *," he muſt have been living in 1704. He had transformed a dead rat into the feigned figure of a dragon, which impoſed upon the learned ſo far, that "ſeveral fine copies of verſes were wrote on "ſo rare a ſubject." BOBART afterwards owned the cheat; but it was preſerved for ſome years, as a maſter-piece of art. There is a print of the elder BOBART, with a diſtich, dated 1675, by *Burghers*; which confirms his German origin; but it is very ſcarce †.

* Part I. Canto ii. l. 314.

† The name of BOBARTIA was given by LINNÆUS to a plant of the *graminaceous* tribe, which he firſt diſcovered in HERMAN's collection of the plants of *Zeylon*.

CHAP.

CHAP. 24.

A short history of the rise and progress of system, method, *or* classification *of plants; from its origin to its revival in* England—*General state of arrangements before* Gesner *and* Cæsalpine —Ray *and* Morison *both laboured in the revival of* method *at the same time—Advantages of* system—*Various methods of* classification *enumerated.*

METHOD.

AGREEABLY to my purpose, I now proceed to give a concise account of the rise and progress of what is understood by *method*, *system*, or *classification* of plants, arising from agreement in the parts of fructification, independent of any association from the *facies externa*, or habit of the plant. To this, I shall add as brief an history of another important discovery, that of the *sexes of plants*; in consequence of which, system itself has been carried to a much higher degree of perfection.

There are no traces of what the moderns

call

call *ſyſtem*, in the writings of the antients; by whom are pre-eminently ſignified, THEOPHRASTUS, DIOSCORIDES, and PLINY. Their knowledge of vegetables was confined to a few that were uſed in medicine, and in the arts and conveniences of life; and in treating on them, their ſubjects are placed in great and inordinate diviſions, without the ſmalleſt approach to what is now meant by *claſſification*.

THEOPHRASTUS treats his ſubject, in general, philoſophically. In his book "*De Cauſis Plantarum*," he conſiders the propagation, culture, qualities, and uſes of Plants in general; but deſcribes very few. In his "*Hiſtoria Plantarum*," in which are deſcribed, or enumerated, about 500 ſpecies, he begins with the organization, the generation, and propagation of Vegetables. He then treats largely, in his *third* and *fourth* books, on Trees. In the *fifth*, on Timber, and the choice of the beſt. In the *ſixth*, on Shrubs, thorny Plants, Roſes, and other ornaments of gardens. In the *ſeventh*, on oleraceous Plants, and wild Plants. In the *eighth*, copiouſly on Grain of all kinds. And

And in the *laſt*, on Gums, Exudations, and the methods of obtaining them.

The object of DIOSCORIDES being ſolely the *Materia Medica*, he diſcuſſes each ſubject ſpecifically, and in a ſeparate chapter, dividing the whole into five books; in which, as far as any order takes place, they arrange into aromatic, alimentary, and medicinal plants. His deſcriptions are taken chiefly from colour, ſize, mode of growing, compariſon of the leaves and roots, with other plants well known, and therefore left undeſcribed. In general they are ſhort, and frequently inſufficient to determine the ſpecies. Hence aroſe the endleſs, and irreconcileable contentions, among the commentators. In this manner he has deſcribed near 700 plants; to which he ſubjoins the virtues and uſes. To DIOSCORIDES all poſterity have appealed as deciſive on the ſubject.

PLINY, who treats of plants from the twelfth to the twenty-ſeventh book, incluſive, of his "Hiſtory," has drawn his reſources principally from Grecian authors. He is the hiſtorian of antient botany, and recites the names of ſeveral hundreds, not mentioned

mentioned by foregoing writers; but many of theſe are unknown. There is no ſcientific order in the diſpoſition of his ſubject; and the great value of PLINY's work conſiſts in having preſerved to us the remains of antient knowledge on the ſubject; and in particular, the application of it to the arts of life, in thoſe remote times.

After the revival of learning in the fifteenth century, the firſt cultivators of botany ſtudied plants more in the writings of theſe fathers, than in the book of nature; and were ſolely anxious about extricating the plants of the *Materia Medica*; ſcarcely adverting to thoſe ſtriking diſcriminations in the general port, mein, or habit, the mode of growing, and other obvious relations, which mark the great natural families in the vegetable kingdom: but were content to arrange them, ſome, according to the alphabetical nomenclature, others, from the ſtructure of the root, the time of flowering, the places of growth, the ſuppoſed qualities, and uſes in medicine; or from other as unſtable diſtinctions. With them, as with the antients, there were nearly as many *genera*

nera as *ſpecies*; and if they gave the ſame common appellation to two, or more plants, they were led to it by ſome rude, external reſemblance; ſuch as, ſize, form of the root, agreement in the colour of the flower; and, in the deſcription of the ſpecies, were frequently ſatisfied with comparing it to another plant well known to themſelves, and therefore left undeſcribed in their writings.

This mode of arrangement, though in a ſomewhat improved ſtate, is exemplified above, in the order obſerved by DODONÆUS; and is ſeen in our old *Engliſh* herbaliſts, GERARD and PARKINSON.

LOBEL, in his *Adverſaria*, 1570, ſeems to have been the firſt, who attempted to diſtribute plants into large families, or claſſes, from the general conſent of habit, or external form, and mode of growing. This he has done in an imperfect ſynoptical way; and ſeveral of his families contain natural orders, or claſſes, nearly entire; but frequently interrupted by great anomalies. His arrangement was not ſufficiently attended to at the time: it was then excellent, and was gradually improved, until we ſee it in its

its laſt, and beſt form, as exhibited by *Caſpar* BAUHINE, in his *Pinax*, 1623; and eſpecially by *John* BAUHINE, in his *Hiſtoria Plantarum Univerſalis*, 1650.

As natural characters aroſe from ſimilarity in the general port, or habit of the plant, and from an obvious agreement in the diſpoſition of the ſtalk, leaves, ſtems, and from that of the flower, fruit, and ſeed; ſo, they at length forced themſelves to obſervation. Thus, the general habit of all *graſſes*; the plants with a *papilionaceous* flower, ſuch as peaſe and vetches; the *ſiliquoſe* plants, ſuch as muſtard, creſſes, turneps, &c.; the *verticillated*, as mint, baum, hyſſop, germander, &c.; the *umbellated* tribe, parſley, carrots, hemlock, angelica; the *cone-bearing trees*; and ſeveral other tribes, were too ſtriking, not to be ſeen even by a ſuperficial obſerver. But, as theſe conſtitute only a part of the whole, ſo no characters were formed for thoſe plants, which the eye could not immediately refer to ſome of theſe claſſes. Still leſs had any generical agreement, ariſing from uniformity in the fructification, been detected. Had all the ſpecies of plants arranged themſelves under

natural

natural classes, a natural method would easily have followed; but the intermediate links, notwithstanding the efforts of the most skilful, are yet unknown. Hence arose the necessity of artificial systems, which are now become but too numerous. Some have imagined, that the more pure any artificial system preserves the natural classes, the greater is its excellence; but experience does not confirm this idea. Those arrangements are found to lead more immediately to the plant sought for, the classes and subdivisions of which are simple, and drawn each uniformly from the same parts of the fructification.

Conrad GESNER, the LINNÆUS of the age in which he lived, is universally agreed to have been the first who suggested this true principle of classical distinction, and generical character, as is manifest from various passages in the *Epistles* of that great man *. He instances the agreement of the *Staphisagria*, with the *Consolida*; the *Scorzonera*, with the *Tragopogon*; the *Molucca*, with the *Lamium*; the *Dulcamara*, with

* Epist. Medicinal. à *Wolphio* ed. p. 113, et passim.

the

the *Solanum*; the *Calceolus*, with the *Orchides*: and he expreſsly ſays, that the character ſhould be formed from the flower, and the ſeed, rather than from the leaves. This was in the year 1565. Other paſſages occur, by which it appears, he had the ſame ideas ſo early as 1559. But, perhaps, there is no proof of the importance he gave to theſe parts, more indubitable, than his having been the firſt who delineated them ſeparately, with the figures of his plants; of which numerous inſtances may be ſeen in the tables publiſhed by SCHMIEDEL.

But GESNER did not live to improve the hints he thus drew from nature; and, what is wonderful, they were neglected by thoſe great luminaries of the ſcience, CLUSIUS, and the BAUHINES. It was reſerved for CÆSALPINUS, a man in whom was united an exquiſite knowledge of plants, with a truly philoſophical genius. He had been the diſciple of GHINUS, and was afterwards phyſician to Pope *Clement* VIII. He deſcribed, with exquiſite ſkill, the plants of his own country, and left an *Herbarium* of

768 ſpecies. He extended GESNER's idea, and commenced the period of ſyſtematic arrangement. In his "*Libri xvi de Plantis*," publiſhed 1583, he has arranged upwards of 800 plants into *claſſes*, founded, after the general diviſion of the trees from herbs, on characters drawn from the fruit particularly, from the number of the capſules and cells; the number, ſhape, and diſpoſition of the ſeeds; and from the ſituation of the *corculum*, *radicle*, or eye of the ſeed, which he raiſed to great eſtimation. The *orders*, or ſubdiviſions, are formed on ſtill more various relations.

Fabius COLUMNA improved this doctrine of *claſſification*, in 1616, by extending it to the formation of *genera*, which CÆSALPINUS had not effected; all his ſpecies being ſeparately deſcribed. COLUMNA, indeed, did not exhibit a ſyſtem; but he ſhewed the way to complete it, by the union of ſpecies under one common name, from ſimilarity in the flower, and fruit; and he invented ſeveral of the terms, now in uſe, to denominate thoſe parts. This noble invention, nevertheleſs, neverthelefs, lay dormant for near a century; and the glory of reviving, and improving it, was referved for *Britain*.

RAY, and MORISON, both laboured in it at the fame time; and with them muft commence the æra of fyftematic botany in *England*. It was an object thought worth contending for, and each of thefe writers had their partizans, who refpectively beftowed the laurel, as they were led by their various motives, or attachments. I fhall not enter into the merits of their claims, further than to obferve, that both feem to have turned their attention to the fubject, nearly about the fame time, and that Mr. RAY had certainly priority in point of publication, if it may be allowed, that the tables which he drew up for Bifhop WILKINS's "Real or Univerfal Character," which was publifhed in 1668, contain the outlines of a fyftem. And, certainly, thefe rudiments, though haftily done, as Mr. RAY confeffes, fufficiently prove that he had beftowed no fmall attention on the fubject. That foreign writers have more commonly attributed to MORISON the revival of method, may have arisen from their being less acquainted with Bishop WILKINS's work, which was extant only in the *English* tongue. Mr. RAY informs us, in the second edition of his *Catalogus Plantarum Angliæ*, that Dr. WILKINS meditated a translation of his "Universal Character" into *Latin*, with figures, for the use of foreigners; and Mr. RAY himself performed it: but the death of this good prelate, in 1672, prevented the completion of the design. He adds, that his Method, in a more elaborate state, had been delivered into the Bishop's hands, for the above-mentioned work.

Dr. MORISON exhibited the outlines of his scheme in the "*Hortus Blesensis*," the year after the publication of the Bishop's book, and exemplified it in his "History of "Plants," in 1680. Mr. RAY did not detail his till the year 1682, in the "*Methodus*," in which he freely acknowledges the assistance he received from CÆSALPINUS, COLUMNA, JUNGIUS, and even from MORISON's work. On the contrary, Dr. MORISON assumes to himself the merit of having drawn all his resources, in the fabrication

brication of his syſtem, wholly from nature, and his own obſervations; preſerving every where the utmoſt ſilence, reſpecting any aſſiſtance derived from former writers. Aſſumptions, which could with difficulty be acceded to, and which drew upon him the cenſures of TOURNEFORT, and other maſters of the ſcience; who were well acquainted with the fountains of knowledge that were then open to him, and the aſſiſtances he muſt have drawn from GESNER, CÆSALPINUS, and others.

At this diſtance of time, and under the preſent enlightened ſtate of ſcience, the ſyſtems of RAY, and of MORISON, muſt not be ſcrupulouſly examined. CÆSALPINUS laid a foundation-ſtone, on which, if our *Britiſh* architects raiſed a *Gothic* ſtructure, their ſucceſſors have improved it to a ſtyle of greater ſymmetry, and elegance.

The introduction of *ſyſtem* was fortunate for ſcience, as it brought with it, by degrees, the eſtabliſhment of generical characters, on a like aſſemblage of eſſential parts in ſeveral ſpecies. As new plants

 were

were daily diſcovered in the old continent, and were pouring in from the new, the nomenclature of botany was in danger of being again overwhelmed, with that chaos in which *Caſpar* BAUHINE found it, when he reduced it into ſome order, by his laborious and incomparable *Pinax*.

Syſtem enabled botaniſts to refer new ſpecies to *genera* already formed, and reſtrained that licence before taken, of giving a new generical appellation to each new plant: for, although in the multitude of methods which followed this diſcovery, plants of the ſame genus, in one ſyſtem, were frequently referable to a different genus in another; yet, with this inconvenience annexed, they were more readily inveſtigated, than under the vague diſtinctions of the older writers.

The reſtoration of *ſyſtem*, was, in the words of LINNÆUS, the beginning of the golden age of botany; and the revival of it having taken place in *England*, preſently raiſed up ſeveral learned men among us, who gave new life and vigour to the whole ſcience. The names of SLOANE, PLUKE-

NET,

NET, SHERARD, and PETIVER, will ever remain illuſtrious in the annals of botanic knowledge.

It alſo turned the attention of the learned on the continent to the ſubject. Rival ſyſtems were ſoon conſtructed; ſome on the *fruit*, as the baſis of the claſſes, in conformity to the ſyſtems of CÆSALPINUS, RAY, and MORISON; and others, on the *flower*. Thus, *Chriſtopher* KNAUT, in 1687, and HERMAN, in 1690, fixed on the *fruit*; whoſe ſyſtems were improved by BOERHAAVE, in 1710.

RIVINUS, in 1690, choſe the *flower* alone; conſidering the *number* and *regularity* of the *petals*, as the baſe of his claſſical characters; and was followed by RUPPIUS in 1718, and LUDWIG in 1737. TOURNEFORT, who elaborated his *method* beyond his predeceſſors, in 1694, choſe the *figure* of the *corolla*, as the principle of *claſſification*; and MAGNOL, in 1720, took the *calyx* alone.

If it ſhould be enquired on this occaſion, in what the *methods* of CÆSALPINUS, RAY, and MORISON, differ from the arrangements uſed before their time, by DODONÆUS, LOBEL, and *John* BAUHINE, ſince thoſe alſo are eſtabliſhed on the habit, and in which many of the natural claſſes are tolerably well preſerved; it may be anſwered in a ſummary way, that habit, even in BAUHINE's order, the moſt perfect of them, is the prevailing principle, without regard to agreement in the parts of fructification, except in thoſe claſſes, where nature has joined both together: this is a difference much more eſſential than may at firſt be apprehended: and, what is ſtill leſs accurate than a regard to habit alone, ſome of their claſſes (if they are worthy of that appellation, no definitions of them being prefixed,) take their name merely from the mode of growing, as, *Scandentes*; from the ſtructure of the leaf, *Nervifoliæ*; *Rotundifoliæ*; *Craſſifoliæ*; place of growth, *Aquaticæ*; and what is ſtill leſs eligible, the aſpect, and ſuppoſed agreement in the qualities, ſuch are, *Malignæ*; *Mollientes*; *Papavera*; under all of which, are promiſcuouſly collected, plants as diſſimilar as poſſible, in the ſtructure of the flower and fruit.

CHAP. 25.

Hiſtory of the diſcovery of the ſexes *of plants—The doctrine of the antients on this head—Their knowledge very limited—The univerſality of this proceſs—The diſcovery of* Millington *and* Grew *—Subſequent writers, who have confirmed or oppoſed the doctrine—Preſent idea of it exhibited.*

SEX OF PLANTS.

TO the revival and eſtabliſhment of *method*, ſucceeded a diſcovery of the higheſt importance to botanical ſcience; I mean, what is, with great juſtice, called analogically, The doctrine of the *ſexes* of plants; or, the knowledge that, throughout the vegetable kingdom, the influence of the duſt of the *antheræ*, upon the *ſtigma*, was neceſſary in order to produce fertile ſeed. By the eſtabliſhment of this fact, not only the phyſiology of vegetables was greatly advanced, but, in the end, practical botany equally improved; ſince, on this foundation has been built that ſyſtem of the great

Swede,

Swede, which is now ſo univerſally followed. Of the riſe and progreſs of this inveſtigation, I proceed to give a conciſe account, before I purſue the ſketch of *Britiſh* authors.

A vague and indeciſive opinion concerning the *ſexes* of plants, prevailed among the antient philoſophers of *Greece*. We are informed by ARISTOTLE, that EMPEDOCLES particularly taught, " that the ſexes " were united in plants." This opinion was a natural conſequence of the doctrine which this philoſopher, in common with ANAXAGORAS, DEMOCRITUS, and PLATO taught, " that plants were ſentient and " animated beings." This idea has met with ingenious advocates among the moderns, who have been induced to favour it, not only from the general analogy exiſting between animals and vegetables, and the difficulty of fixing the limits between them, but from the more ſtriking inſtances of apparent irritability, and obedience to the action of certain *ſtimuli:* ſuch are, the general affection plants have for light; the rotatory motion of many towards the ſun; the faculty of others in cloſing the leaves

at

at night, called, not unaptly, the *ſleep of plants*; and the opening and ſhutting of many flowers, at ſtated times, with equal propriety denominated *vigiliæ florum*; the riſing of the flower of aquatic plants out of the water, every morning during the ſtate of floreſcence, as inſtanced in the *Nymphæa*, and ſtill more ſignally in the *Valliſneria*. To theſe may be added the more remarkable examples in the *Mimoſa*, and *Oxalis ſenſitiva*, in the *Dionæa muſcipula*, the *Droſera* and the *Hedyſarum gyrans*, and finally, in the exquiſite irritability of the *ſtamina*, and *antheræ*, in various ſpecies. EMPEDOCLES, nevertherleſs, though he maintained the doctrine of the *ſexes*, does not attempt to confirm it by any facts, or reaſonings deduced from the knowledge of the uſes of the ſeparate parts in flowers, but from analogical deduction, founded merely on his general doctrine.

ARISTOTLE, or rather the author of the *Books on Plants*, which bear his name, combats the opinions of EMPEDOCLES, and his followers, reſpecting the ſentient and animated principle in vegetables; yet it is evident

dent he had himſelf no deciſive ideas, or ſpecific knowledge, drawn from nature, relating to the *ſex* of plants. He placed it, in ſome inſtances, in the different habit alone, or in other diſcriminations foreign to the conſideration of the flower; and, though he ſhews an inaccurate knowledge of the particular circumſtances of the palm, and the fig-tree, yet he denies, in another place, that either of them produce flowers.

This imperfect idea of the *ſex* of flowers, in the *Date*, and even in the *Fig-tree*, is of high antiquity; being recorded by HERODOTUS, THEOPHRASTUS, and PLINY. The neceſſity which the antient cultivators of the *Date-tree* were under, of promoting the action of the male-flowers on the female, which operation held alſo in ſome meaſure in the *Fig-tree*, the *Piſtachia*, and the *Maſtic*, would almoſt neceſſarily ſuggeſt the application of this analogy with the animal kingdom. Neverthleſs, although the fact was thus obtruded on their ſenſes, inattentive to the ſtructure of flowers, and ignorant of the offices of the ſeveral parts, they remained unacquainted with the true operations

rations of nature in this phænomenon, though daily preſent to their obſervation.

The antient fathers of botany, and particularly DIOSCORIDES, it is true, applied the diſtinction of male and female to many other plants; but it was entirely without regard to true analogy, or diſcrimination of functions in the flower. It was frequently applied to ſuch as carry all the parts of the flower within the ſame *calyx*, or on the ſame ſtalk; on account of ſtature; greater degree of fertility; or other marks unconnected with the fructification. In the *diœcious*, or ſuch as have the *ſtamina*, and *piſtils*, on ſeparate plants of the ſame ſpecies, the real male plant was, in ſome caſes, denominated the female; of which the *Mercurialis* may be mentioned as on inſtance, among ſeveral others.

Excluſive of a numerous ſet of plants, in which the *ſtamina* and *piſtils* are ſeparately placed, either on different parts of the ſame individual, or on different plants of the ſame ſpecies, conſtituting the *Monœcious* and *Diœcious* claſſes of LINNÆUS, the following

following *genera*, from other tribes, as recited below*, contain ſpecies to which DIOSCORIDES has applied the diſtinction of male and female, from circumſtances having no analogy with thoſe of the *Date-tree*.

This doctrine of the ſexual analogy between plants and animals, made but little progreſs with the *literati* in botany, upon the revival of ſcience; ſince the firſt of thoſe who mention it, is CÆSALPINUS. This critical and learned author notices male and female plants in the *Oxycedrus*, *Taxus*,

*

Arundo	Mandragora
Anagallis	Pæonia
Ariſtolochia	Polygonum
Ciſtus	Tithymalus
Filix	Verbaſcum, &c. &c.

To which have been added, by others,

Abrotanum	Nicotiana
Abies	Orchis
Amaranthus	Pulegium
Balſamina	Quercus
Caltha	Symphytum
Cornus	Tilia
Criſta Galli	Veronica,
Ferula	&c. &c.

Taxus, *Mercurialis*, *Urtica*, and *Cannabis*; of which he ſays, the barren plants are males, and the fertile females; adding, that the latter, as is obſerved in the *Date-tree*, becomes more fruitful by being planted near the males; from thence receiving a genial *effluvium*, which excites a ſtronger fertility. From this obſervation, it may almoſt be inferred, that he had inſtituted experiments on ſome of theſe kinds; but we do not find that he carried the idea beyond the above-mentioned ſpecies, to vegetables in general.

Adam ZALUZIANSKY, a *Poliſh* writer in 1592, is ſaid, by ſome, to have diſtinguiſhed the *ſexes* of plants. I have not ſeen his book; but, from what is found relating to his opinion in other writers, I conjecture that his obſervations, if not wholly taken from CÆSALPINUS, do not exhibit any original matter on this ſubject. In fact, no further progreſs was made for near an hundred years after this time; and the honour of the diſcovery, "that this ſexual pro-
"ceſs was univerſal in the vegetable king-
"dom, and that the duſt of the *antheræ*
"was

" was endowed with an impregnating " power," is due to *England*.

Whether the true idea of this proceſs originated with Sir *Thomas* MILLINGTON, to whom it has been aſcribed, may juſtly admit of a doubt; ſince Sir *Thomas* has left no written teſtimony on the ſubject; and Dr. GREW's mention of him does not imply that he actually received the idea from him. Add to this, that Mr. RAY, in the ſummary view of all GREW's diſcoveries, which he has prefixed to his " Hiſtory of " Plants," does not once mention Sir *Thomas* MILLINGTON's name. Intereſted as we muſt ſuppoſe Mr. RAY to have been, in every diſcovery relating to vegetables, and candid as he was in his general conduct to the learned, it is not likely that he ſhould have failed, in this inſtance, to render praiſe where it was ſo juſtly due. When we further recollect, that Dr. GREW had been ſome years engaged in thoſe microſcopical experiments, on the anatomy of plants, which have rendered his name eſtimable with all poſterity, that whilſt he was thus employed in ſtudying ſo intimately the

the organization of vegetables, and had obſerved, that in whatſoever parts the flower might be deficient, the attire, (or *ſtamina*, and *apices*) is ever preſent, it is not ſtrange that the true idea of its uſe ſhould have been ſuggeſted to him.

Dr. GREW laid his opinion before the Royal Society, in a lecture on the anatomy of flowers, read Nov. 6, 1676; in which he maintained, "That the primary and
" chief uſe (of the duſt of the *apices*) is
" ſuch as has reſpect to the plant itſelf,
" and ſo appears to be very great and ne-
" ceſſary: becauſe even thoſe plants which
" have no flower, or foliature, are yet ſome
" way or other *attired*, ſo that it ſeems to
" perform its ſervice to the ſeed as the fo-
" liature to the fruit. In diſcourſe hereof
" with our learned *Savilian* profeſſor, Sir
" *Thomas* MILLINGTON, he told me, that
" he conceived that the *attire* doth ſerve
" as the male for the generation of the
" ſeed. I immediately replied, that I was
" of the ſame opinion, gave him ſome rea-
" ſons for it, and anſwered ſome objections
" that might oppoſe them." He then ex-

 plains

plains himſelf farther, and advances, that this fœcundating power was not effected by the actual admiſſion of the *farina* into the ſeed-veſſel, but by means "of ſubtle and "vivific effluvia."

Mr. RAY admitted the opinion of Dr. GREW, but, at firſt, with all that caution which becomes a philoſopher; as appears in his "*Hiſtoria Plantarum*," vol. i. p. 18. *Nos ut veriſimilem tantum admittimus.* He aſſents to it with leſs reſerve in his "*Synopſis Stirpium Britannicarum*," edit. 1. 1690, p. 28; and in the preface to his "*Sylloge Stirpium Europæarum*," publiſhed in 1694, we find him producing his reaſons for the truth of it, and yielding his full approbation to it.

In 1695, *Rudolph Jacob* CAMERARIUS, profeſſor of botany and phyſic at *Tubingen*, in his "*Epiſtola de Sexu Plantarum*," appears among the early advocates for this analogy; and, being convinced by the arguments of GREW and RAY, ſeems to have been the firſt who gave ſtability to the whole by experiments. Theſe he made on *Maize*, the *Mulberry*, the *Ricinus*, and the *Mercurialis*; the three firſt of which he deprived

deprived of the ſtaminiferous flowers, and the laſt he ſeparated far from the female, and found, in all, that the fruit did not ripen. CAMERARIUS, however, very fairly produces alſo, ſome objections againſt the doctrine, founded on experiments, which at this day have little weight, ſince they were made on plants of the *Cryptogamous*, or *Dioecious* claſſes; in the laſt of which, it is now known, that ſometimes a flower or two of a different ſex, may be found intermixed with others.

In 1703, Mr. *Samuel* MORLAND, deſirous, as it ſhould ſeem, of extending the *Lewenhoekian* ſyſtem of generation into the vegetable kingdom, produced a paper before the *Royal Society*, in which he advances—that the *farina* is a *congeries* of ſeminal plants, one of which muſt be conveyed through the ſtyle into every *ovum*, or ſeed, before it can become prolific. Mr. MORLAND's hypotheſis tended to confirm the general doctrine by exciting curioſity on the ſubject, at a time when *Lewenhoek*'s theory was popular; but was not admiſſible in itſelf, ſince few ſtyles are hollow, or, if perceptibly tubular, not pervious enough to admit particles of the ufual magnitude of the *farina*.

After this time, feveral of the learned on the continent entered into refearches on this fubject. *M.* GEOFFROY, in 1711, in a paper read before the *Royal Academy of Sciences*, after having formed a theory by conciliating GREW's and MORLAND's into one, concludes by afferting—that the germ is never to be feen in the feed, till the *farina* is fhed; and that if the plant is deprived of the *ftamina*, before this duft is fallen, the feed will either not ripen, or will not prove fertile.

It is matter of furprize, that the illuftrious TOURNEFORT fhould wholly reject the doctrine of the *fexes* of plants. So far even from acknowledging this function of the *farina*, that he held it to be excrementitious. See *Ifagoge in Rem Herbariam*, p. 70.

Julius PONTEDERA, a ftrenuous follower of TOURNEFORT, a noble *Italian* of *Pifa*, illuftrious for his knowledge of the antient languages, and antiquities of *Italy*, and not lefs celebrated for botanical knowledge and literature, combats alſo the notion of this analogy, and uſes of the *ſtamina*, through the whole ſecond book of his "*Anthologia.*" In the end he rejects the ſexual analogy, and conſiders it as entirely chimerical. But finding all flowers furniſhed with a ſtyle, or tube, he advances, that it ſerves to convey the air to the fruit, by which, an inteſtine and fertilizing motion is excited in the ſeed, or ovary.

In 1718, Monſ. VAILLANT publiſhed "*Sermo de Structura Florum, horum Differentia, uſuque Partium*;" which had been read the year before, at the opening of the Royal Garden. In this diſcourſe, he deſcribes the burſting of the *antheræ*, in a ſtyle too florid for philoſophical narration. He relates ſeveral of his own diſcoveries on the nature of the *farina*, and the exploding power of the *antheræ*, and concludes with aſſenting entirely to Dr. GREW's ſentiment, (though without naming him), that impregnation is performed by means of a ſubtle *aura*, and not by the tranſmiſſion of the duſt through the ſtyle, alledging againſt it thoſe reaſons I have mentioned, in ſpeaking of MORLAND's opinion.

In *England*, about the ſame time, Dr. *Patrick* BLAIR, by his " Botanick Eſſays," contributed greatly to extend the knowledge, and confirm the truth of this ſubject. BRADLEY, FAIRCHILD, MILLER, and others, aſſiſted in the ſame deſign; and, ſince that period, I believe it has met with few oppoſers. One of the moſt formidable was the late learned Dr. *Alſton*, profeſſor of botany at *Edinburgh*, from whoſe laboured diſquiſition, the adverſaries to this opinion of the *ſex* of flowers, may furniſh themſelves with the moſt cogent arguments, that an intimate knowledge of the ſubject hath enabled a very diligent and learned writer to produce.

The more recent experiments made by the Abbe SPALANZANI, with a direct view to impugn this doctrine, do not appear to have been conducted with that degree of ſkill, and accuracy, which is ſufficient to outweigh the numerous train that may be thrown into the oppoſite ſcale. Even ſome of the Abbe's own experiments ſeem rather to ſtrengthen the opinion he means to overthrow.

Having traced the hiſtory of this important proceſs in the economy of vegetables, to the time of LINNÆUS, I judge it will be unneceſſary, to accompany the reader through a particular detail of authors below this period. In 1732, LINNÆUS founded his ſyſtem on this doctrine; and the additional arguments, and experiments, produced by himſelf, his pupils, and followers, have eſtabliſhed the truth of it, to the compleat ſatisfaction of impartial enquirers. Thoſe, however, who wiſh to peruſe the moſt perfect ſummary of all the arguments, and experiments, in favour of this analogy, are referred to the "*Sponſalia Plantarum*," written in the year 1746, and printed in the firſt volume of the "*Amœnitates Academicæ*," and to the "Diſſertation on the Sexes "of Plants," written by LINNÆUS in 1760, which obtained the premium of the Academy of *Peterſburgh*, and has lately been tranſlated into *Engliſh* by the ingenious and learned poſſeſſor of the *Linnæan* collec-

tion.

tion. To which may be added, the writings of KOELRUTER, in the succeeding year, which have not a little tended to confirm the subject in question.

It would be unjust to the memory of Dr. GREW, to conclude this history, without remarking, that the result of the latest, and best experiments, have confirmed his idea, "that the *farina* itself is not carried to the "*rudiment* of the seed," but, that fœcundation is effected by the effluvia. This will appear, by citing the summary view of the doctrine, as exhibited by LINNÆUS himself, in the Dissertation above mentioned.

"While plants are in flower, the *pollen* "falls from the *antheræ*, and is dispersed "abroad. At the same time that the *pollen* "is scattered, the *stigma* is then in its "highest vigour, and for a portion of the "day at least is moistened with a fine dew. "The *pollen* easily finds access to the *stigma*, where it not only adheres by means "of the dew of the part, but the moisture "occasions its bursting, by which means "its contents are discharged. What issued

"from

"from it being mixed with the *ſluid* of "the *ſtigma*, is conveyed to the *rudiments* "of the *ſeed.*"

I remark before I conclude, that, how juſt ſoever it may have been in a philoſophical view, to conſider the *ſtamina* and *piſtils*, as anſwering to the reſpective functions of *ſex* in the animal kingdom, it ſhould not have been forgotten, that in animals, this proceſs is voluntary; but that in vegetables, notwithſtanding all that the ingenuity of the antients and moderns have urged in defence of the ſentient principle, we are not *yet* juſtified in referring this proceſs to any other than what we are accuſtomed to call a mechanical cauſe.

The principle of this it will not be expected that I ſhould explain. It may be conjectured, that after a perfect elaboration of the juices in the *antheræ* and *ſtigmata*, ſome ſpecies of attraction takes place between them, perhaps of the electrical kind, ſomewhat like this having been manifeſted in the flaſhings obſervable in ſome flowers in the evenings. The reader will eaſily perceive, that I refer to the appearance firſt ſeen

ſeen in the *Indian Creſſes, (Tropæolum majus)* by *Elizabeth Chriſtina*, the daughter of LINNÆUS, as related in the *Swediſh Acts* in 1762, and ſince confirmed in the *Garden Marigold (Calendela officinalis)*, the *Orange*, or *bulbiferous Lily (Lilium bulbiferum)*, and the *African Marigold (Tagetes patula et erecta)*, by the obſervations of *M.* HAGGREN. And, as in the univerſe at large, the phænomena of electricity are ſenſibly manifeſted to us by particular modifications of the principle occaſionally excited, although unqueſtionably ever active, ſo, poſſibly, the ſame principle may prevail through the whole vegetable creation in the proceſs above mentioned, though unobſerved hitherto, except in theſe inſtances. Be this as it may; that general decorum, which is due to philoſophical ſubjects, ought to have reſtrained that reprehenſible language uſed by *Vaillant*, and ſome other writers on this ſubject, and even by LINNÆUS himſelf, which has juſtly diſguſted many readers, and prejudiced the inſtruction they meant to convey.

CHAP. 26.

Willisel—*Collects plants for* Merret, Morison, Ray, *and* Sherard—*His Notices on the* Misseltoe.

Plott—*Anecdotes of—His Natural history of* Oxfordshire *and* Staffordshire.

Natural history of counties—Plott *the first writer*—Leigh's Lancashire—Robinson's Westmorland — Moreton's Northamptonshire — Borlace's Cornwall—Wallis's Northumberland.

Wheler—*Anecdotes of—Journey into* Greece—*Introduced some new plants into* England.

WILLISEL.

IT is not to the sons of erudition alone, that botany is indebted for all its discoveries, and improvements. The love of plants has, not unfrequently, seized, with uncommon ardour, the minds of many, on whom the light of learning had not shed its influence; and spurred them on, in the pursuit of this knowledge, to attainments that have been highly beneficial to the science.

ence. From ſuch, let not the pride of learning withhold that praiſe which is ſo juſtly due. One of the moſt remarkable inſtances of this kind, is well known to thoſe who are converſant with the writings of MERRET, RAY, and MORISON; and I feel regret at not being able to commemorate the name of *Thomas* WILLISEL, with ſome of the circumſtances of his life; ſince I am uninformed of the time, and place, both of his birth, and of his death. This induſtrious man ſeems to have devoted much of his life to the inveſtigation of *Engliſh* plants; and, as he lived at a time when *Britiſh* botany was yet imperfect, he added largely to the ſtock of new diſcoveries. He was employed by Dr. MORISON, ſoon after his eſtabliſhment at *Oxford*, to collect rare *Engliſh* plants; and Dr. MERRET informs us, as hath been noticed, that he travelled five ſummers at his expence, into the different parts of *England*, to make collections for his "*Pinax*;" which appears to have been greatly enriched with many of the moſt rare ſpecies, by the labours of WILLISEL.

I believe

I believe he was once sent into *Ireland* by Dr. SHERARD. Mr. RAY was benefited by his researches; and, if I do not mistake, he accompanied that celebrated naturalist in one of his tours. The emolument arising from these employments was probably among the principal means of his subsistence.

His knowledge was not confined to the vegetable kingdom; since Mr. RAY informs us, that "he was employed by the "*Royal Society* in the search of natural ra-"rities, both animals, plants, and mine-"rals; for which purposes he was the fit-"test man in *England*, both for his skill "and industry."

In the letters of Mr. RAY, there occurs an observation made by WILLISEL, of the various trees on which he had found the *Misseltoe* growing. I enumerate them below *.

* Oak. Ash. Lime. Hasel. Willow. White Beam. Purging Thorn. Quicken Tree. Apple Tree. Crab Tree. White Thorn.

PLOTT.

Dr. *Robert* PLOTT, eminent for being the firſt who ſketched out a plan for a natural hiſtory of *England*, by exemplifying it in that of *Oxfordſhire* and *Staffordſhire*, although not profeſſedly a writer in the botanic line, cannot be omitted in a work of this kind.

He was born at *Borden*, near *Sittingborne*, in *Kent*, and educated at *Wye*, in the ſame county; entered a ſtudent in *Magdalen Hall*, in 1657; and, in 1671, took the degree of doctor of laws. He became fellow of the *Royal Society*, and was made one of the ſecretaries in 1682. In the ſame year he was conſtituted the firſt keeper of the *Aſhmolean* Muſeum, and profeſſor of chymiſtry: all which places he kept till 1690; having alſo, in 1687, been appointed *Mowbray* herald extraordinary, and regiſter to the earl marſhal, or court of honour, then newly revived, after having lain dormant from the year 1641. He died April 30, 1696. There is a whole length portrait of him,

him, the laſt of the right hand group, in the *Oxford* Almanack for the year 1749.

Dr. PLOTT was a man of various erudition, but is at this time beſt known for his natural hiſtories of *Oxfordſhire*, and of *Staffordſhire*. The firſt of theſe was publiſhed in 1677, in folio; and again in 1705, with the author's corrections and additions, by his ſon-in-law, Mr. *Burman*, vicar of *Newington*, in *Kent*. The natural hiſtory of *Staffordſhire*, in 1679, in folio, and reprinted in 1686. In each of theſe volumes, he records the rare plants of the county, deſcribes the dubious ones, and ſuch as he took ſor nondeſcripts, and figures ſeveral of them. To theſe works the *Engliſh* botaniſt owes the firſt knowledge of ſome *Engliſh* plants; and this circumſtance juſtly entitles him to a place in this work *. He conducted the publication of the *Philoſophical Tranſactions* during part of his ſecretaryſhip to the Society, and wrote the following papers:

* It is amuſing to remark the price of literature a century ago The ſubſcription for PLOTT's *Staffordſhire* was, a penny a ſheet, a penny a plate, and ſix pence the map.

A Paper

A Paper on the Formation of Salt and Sand from Brine of the Pits in *Staffordshire*. Printed in N° 145.

On Perpetual Lamps, in imitation of the ſepulchral lamps of the antients. N° 166.

On the Incombuſtible Cloth made of the *Aſbeſtos*. Ib.

A Hiſtory and Regiſter of the Weather at *Oxford* during the year 1684. N° 169.

On the Black Lead of *Cumberland*. N° 240.

On the beſt Time for felling Timber, which, with the antients, he adviſes to be performed in the Autumn.

On an *Iriſh* Giant, nineteen years of age, and meaſuring ſeven feet ſix inches in height. N° 240.

A Catalogue of Electrical Bodies. N° 245.

NATURAL HISTORY OF COUNTIES.

I have before obſerved, that Dr. PLOTT was the firſt author of a ſeparate volume on Provincial Natural Hiſtory; in which, it is but juſtice to add, that, with due allowance for the time when he wrote, he has not been

been excelled by any ſubſequent writer. It were to be wiſhed, that more examples of the like kind might be adduced; but there are few exactly of the ſame ſcope. After Biſhop GIBSON, in his edition of CAMDEN, printed in 1695, had inſerted the provincial liſts of plants drawn up by Mr. RAY, ſeveral writers of county hiſtories have, either from their own knowledge of the ſubject, or by the aid of friends, inſerted catalogues of the *more rare* plants in their reſpective works. As theſe form, in an eſpecial manner, a part of *Engliſh* botany, it is incumbent upon me to enumerate them.

The firſt after CAMDEN, is "The Na-" tural Hiſtory of *Lancaſhire, Cheſhire,* and "the Peak in *Derbyſhire.*" Oxford, 1700. fol. By *Charles* LEIGH, M. D. The author takes into his catalogue the maritime plants, with the others, and briefly recites the virtues, and the medicinal claſſes, to which the ſubjects belong. He ſubjoins his conjectures on the food of vegetables, and conteſts the opinion of Dr. WOOD-

WARD, that plants are nouriſhed by the earthy principle alone.

" An Eſſay towards a Natural Hiſtory " of *Weſtmorland* and *Cumberland*, wherein " an account is given of their ſeveral mi- " neral and ſurface productions." By *Thomas* ROBINSON, rector of *Ouſby*, in *Cumberland*. 1709. 8°. The ſcope of this volume principally takes in the foſſils of theſe northern counties. The author has been mentioned before, as a correſpondent of Mr. RAY. He here enumerates profeſſedly the plants not mentioned in the *Synopſis* of that author, amounting to about twenty; of which, however, ſome were only varieties.

" The Natural Hiſtory of *Northampton-* " *ſhire*, with ſome account of the Antiqui- " ties." By *John* MORETON, A.M. F.R.S. rector of *Oxendon*, in the ſame county. Lond. 1712. fol. This is a work of merit. In the liſt of plants, ſeveral occur additional to thoſe noticed by RAY; even ſome of the moſſes are not forgotten. The author treats largely on figured foſſils, of which

which his book contains many elegant plates.

Of "the Natural Hiſtory and Antiqui-"ties of *Surrey*, begun in the year 1673, "by *John* AUBREY, Eſq. F. R. S.; pub-"liſhed by Dr. RAWLINSON, in 5 vol. 8°. "Lond. 1719;" I can only recite the title.

In the "Natural Hiſtory of *Cornwall*," by *William* BORLACE, A.M. F.R.S. Oxford, 1758, we meet with a very brief liſt, containing about thirty-eight land plants, and twenty *fuci*, with ſome ſcattered remarks on the qualities and uſes. Among the rare plants are the Verticillate Knotgraſs, the Roman Nettle, the Gunhilly Heath, and the Corniſh Pennywort; of which laſt there is a very indifferent figure in tab. 29. f. 6. Under the article Sun-dew, (*Droſera*) there is a curious and intereſting obſervation made by Dr. BORLACE, in which he aſſerts, that the well-known pernicious quality of that vegetable, in producing the rot among ſheep, where it abounds, does not ariſe from any cauſtic power in the vegetable, but from an inſect, which lays its eggs, and feeds on the

 plant.

plant. From his account, this insect appears to be the *Dropsy Worm* of Dr. TYSON, or the *Hydra Hydatula* of LINNÆUS.

"The Natural History and Antiquities "of *Northumberland*, and of the North Bi-"shopric of *Durham*, lying between the "*Tyne* and *Tweed*." By *John* WALLIS, M.A. 2 vol. 4°. Lond. 1769. The eighth chapter of the first volume treats on the vegetable productions of this tract, with the various medicinal and œconomical uses.

In the "History and Antiquities of the "Counties of *Westmorland* and *Cumber-*"*land*," by *Joseph* NICHOLSON, Esq and *Richard* BURN, LL.D. 2 vol. 4°. 1777, the reader will meet with some observations on the natural history interspersed; but the botanist will find but little interesting in his way.

From CAMDEN, from these histories, and other resources, Professor MARTYN has compiled an abridged list of all the rare plants, digested in the order of the counties, which is intended for the use of the travelling

travelling botanift See the "*Plantæ Cantabrigienſes.*" Lond. 1763; from p. 44—144.

WHELER.

As I do not ſtrictly confine myſelf to ſuch writers, as have diſtinguiſhed themſelves by their diſcoveries in the indigenous botany of *Britain*, alone, I cannot therefore omit to mention ſo eminent a man as Sir *George* WHELER. He was the ſon of Col. WHELER, of *Charing*, in *Kent*; and was born in 1650, at *Breda*, his parents being there in exile with the royal family. At the age of ſeventeen, he became a commoner of *Lincoln College*, *Oxford*; and, before he took any degree, went on his travels. He ſpent near two years in *France* and *Italy*; and, in 1675, travelled into *Greece* and *Aſia Minor*; from whence he returned in *November* 1676. He was knighted before he took his maſter of arts degree, which was conferred upon him in 1683, in conſideration of his learning, and in return for a preſent of antiquities collected in his travels. He afterwards took ſome

ſome valuable preferments in the church; was created doctor of divinity in 1702; and died Feb. 18, 1724.

In 1682, was publiſhed, " A Journey " into *Greece*, by *George* WHELER, Eſq. in " company of Dr. SPON, of *Lyons*; in ſix " books; with four tables of coins, and " many other ſculptures." Fol. pp. 483.

Theſe gentlemen travelled with PAUSANIAS in their hands, by whoſe means they corrected, and explained, ſeveral of the antiquities and traditions of *Greece*. The primary objects of theſe learned travellers were, to copy the inſcriptions, and deſcribe the antiquities and coins of *Greece* and *Aſia Minor*, and particularly of *Athens*, where they ſojourned a month. Theſe travels are highly valued for their authenticity, and are replete with ſound and inſtructive erudition to the medalliſt and antiquary.

Mr. WHELER appears, on all occaſions, to have been attentive to the natural hiſtory of *Greece*, and particularly to the plants, of which he enumerates ſeveral hundreds in this volume, and gives the engravings of ſome. Theſe catalogues ſufficiently evince

his

his knowledge of the botany of his time. He brought from the Eaſt ſeveral which had not been cultivated in *Britain* before. Among theſe, the *Hypericum olympicum* (St. John's Wort of Olympus) is a well-known plant, introduced by this learned traveller. RAY, MORISON, and PLUKENET, all acknowledge their obligations for curious plants received from him.

After Sir *George* WHELER entered into the church, he publiſhed "An Account of "the Churches and Places of Aſſembly "of the Primitive Chriſtians; from the "Churches of *Tyre*, *Jeruſalem*, and *Conſtantinople*, deſcribed by *Euſebius*, and "ocular Obſervations of ſeveral very an-"tient Edifices yet extant in thoſe Parts: "with a ſeaſonable Application." Lond. 1689.

The Rev. *Granville* WHELER, of *Otterden Place*, *Kent*, and rector of *Leak*, in *Nottinghamſhire*, who died in 1770, was the third ſon of Sir *George* WHELER, and became his heir. He diſtinguiſhed himſelf as a gentleman of ſcience, and a polite ſcholar. He was the friend and patron of Mr. *Stephen*

GRAY; who, jointly with him, contributed to revive the ſtudy of electricity in *England*. Let me be allowed to add, that I wiſh to mention the name of this gentleman with gratitude, from the recollection of that encouragement which I perſonally received from him in my purſuits of natural hiſtory, at a very early period of life; and which was of ſuch a nature, as ſeldom fails to animate the minds of the young, to exertion and improvement.

END OF THE FIRST VOLUME.

Printed in the United States
By Bookmasters